Once Can Be Enough

Allan Franklin · Ronald Laymon

Once Can Be Enough

Decisive Experiments, No Replication Required

Allan Franklin
University of Colorado
Boulder, CO, USA

Ronald Laymon
The Ohio State University
Columbus, OH, USA

ISBN 978-3-030-62567-2 ISBN 978-3-030-62565-8 (eBook)
https://doi.org/10.1007/978-3-030-62565-8

This Springer imprint is published by the registered company Springer Nature Switzerland AG
The registered company address is: Gewerbestrasse 11, 6330 Cham, Switzerland

Acknowledgements

This book has been a collaboration, so anyone who has helped one of us has, in fact, helped both of us. With that in mind, we would like to express our gratitude to the following people. As always members of the high-energy physics group at the University of Colorado have been available to answer questions, discuss technical issues, and have provided important material. They are John Cumalat, Shanta DeAlwis, Oliver DeWolfe, Bill Ford, Alysia Marino, Keith Ulmer, Steve Wagner, and Eric Zimmerman. For many years, Cynthia Betts has provided support and encouragement.

Contents

About the Authors

Allan Franklin is Professor of physics emeritus at the University of Colorado. He began his career as an experimental high-energy physicist and later changed his research area to history and philosophy of science, particularly on the roles of experiment. In 2016, Franklin received the Abraham Pais Prize for History of Physics from the American Physical Society. He is the author of eleven books including most recently *Shifting Standards: Experiments in Particle Physics in the Twentieth Century*, *What Makes a Good Experiment?: Reasons and Roles in Science*, and *Is It the Same Result? Replication in Physics.* He is co-author, with Ronald Laymon, of *Measuring Nothing, Repeatedly: Null Experiments in Physics*.

Ronald Laymon is Professor of philosophy emeritus at the Ohio State University, where he specialized in the history and philosophy of science. He has published widely, was the recipient of multiple National Science Foundation research grants, and was a fellow at the Center for the Philosophy of Science at the University of Pittsburgh, and a resident scholar at the Rockefeller Foundation's villa in Bellagio. In 1995, he took advantage of an early retirement option and completed a law degree at the University of Chicago School of Law in 1997. He then went on to practice large-scale commercial litigation at the Jones Day law firm where he had the good fortune to serve as second chair on a case before the Supreme Court of the United States. Now retired from the full-time practice of law, Laymon does consulting work for a biotech, intellectual property firm that facilitates the open-source creation of therapeutic technologies. Retirement has also made it possible for Laymon to resurrect his interest and earlier work in the history and philosophy of science. He is co-author, with Allan Franklin, of *Measuring Nothing, Repeatedly: Null Experiments in Physics*.

Chapter 1
Introduction

Replication has often been claimed to be the Gold Standard for scientific objectivity. In fact, unquestionably so.

> Replication – the confirmation of results and conclusions from one study obtained independently in another is considered the scientific gold standard. Replication – the confirmation of results and conclusions from one study obtained independently in another is considered the scientific gold standard.
> (Jasny et al. 2011).
>
> Replications are considered by many to be the "scientific gold standard" because they are essential for the ability of a scientific field to self-correct, which is one of the hallmarks of science (Kepes and McDaniel 2013, p. 261, citations omitted).
>
> Reproducibility is the gold standard for scientific research. The legitimacy of published work depends on whether we can replicate the analysis and reach the same results (Janz 2015, p. 1).

The Gold Standard has been seen as especially and distressingly relevant in the social sciences because of many well publicized replication failures. Thus, there has been extensive commentary on the enforcement of the standard and ensuring that experimental results are more likely to pass muster (see, e.g., National Academies of Sciences 2019).

But even if acceptance of the Gold Standard is deemed axiomatic, we can still ask why this exalted status? John Norton has provided a well-considered starting place:

> The general idea is simple and instantly compelling. If an experimental result has succeeded in revealing a real process or effect, then that success should be replicated when the experiment is done again, whether it is done by the same experimenter in the same lab. .. or by others, elsewhere, using equivalent procedures. ... (Norton 2015, p. 2).

Succinct and indeed "compelling"—at least at first glance. But given a few moments thought many ambiguities become evident. Begin with the question: What is it exactly that's being replicated? Construed narrowly, an "experimental result" is what is sometimes described as the "raw" data. But, of course, such data is never entirely theory free insofar as the theoretical underpinning of the instrumentation is assumed. The range here is enormous: rulers, thermometers, timing devices,

A. Franklin and R. Laymon, *Once Can Be Enough*,
https://doi.org/10.1007/978-3-030-62565-8_1

radioactive labeling of DNA segments, cloud chambers and much more. Norton, though, requires more than just raw data results. Such results must have "succeeded in revealing a real process or effect" before the replication requirement applies. Reading between the lines, what's at issue is whether the experimenters involved have claimed that their theory-mediated raw data succeeds in revealing a real process or effect. Expressed otherwise, what's required is not just raw data but raw data that is claimed to reveal some underlying theoretically significant process or effect.

Putting aside for now the ambiguities and vagueness involved, we ask of what value is replication in such cases? In other words, why is replication the Gold Standard for experimental research? Presumably, what's involved here is the relevance of having been replicated for confirming that an experiment has been properly performed. For example, that confounding causes were adequately accounted for. Or that those conducting the experiment were not biased in their selection of the raw data suitable for further analysis. The idea then is that if there exist such confounding causes or bias then repetition of the experiment (presumably under yet to be specified different conditions) will return conflicting results that reflect these experimental defects.

Fair enough. But now we ask whether such experimental duplication is an effective way of achieving this end? There is an immediate problem, and obviously so, because duplication by itself may serve only to duplicate whatever defects that might have existed in the original experiment. This suggests that any application of the Gold Standard must involve some specification of possible defects about which there is reason to be concerned and that any replication be made in a way that maximizes the chance of obtaining a different and revealing result.

It is important at this stage to ask whether the Gold Standard is to be understood as requiring as a necessary condition for experimental respectability that an experiment be in fact replicated. If that be the intent, it is clearly mistaken. And this for the simple reason that there exist a great many other strategies—to be discussed shortly—that can be and were employed as evidence of experimental respectability. Thus, the Gold Standard should be read as requiring only that *if* the experiment were performed again, then the same experimental success would result. But the truth of this conditional could be shown or supported by successful applications of the other strategies for validating experimental results. In short, the Gold Standard is just one among many strategies for such validation. The bottom line then is that the real question is what are the efficacious and efficient strategies for validating experimental results?

In an initial effort to specify the range of strategies that can, and have been, legitimately used to argue for the credibility of an experimental result, Allan Franklin has proposed the following possibilities—what he has referred to as "an epistemology of experiment" (Franklin 2016, p. 306). The first three possibilities deal with various forms of manipulation of the experimental apparatus and test conditions in order to determine (1) whether the experiment has sufficient sensitivity to register the anticipated effect; (2) to what degree the apparatus is sensitive to the variation of anticipated confounding effects; and (3) whether known phenomena can be reproduced as a sort of precondition for the effective functioning of the apparatus. Franklin then adds, as a means of avoiding possible experimenter bias, (4) the use of various forms of "blind" analysis by setting the selection criteria for "good" data independent of the final result. There is also the more general (Sherlock Holmes) strategy of (5) arguing that there is no plausible malfunction of the apparatus, or background effect, that would otherwise account for the observations.

The next strategy involves an application of the truism that if the prediction of a independently, well-corroborated theory fails, that is a reason for suspicion of the result and replication of the experiment. But if an experiment is found confirming, the other side of the truism applies, namely, when a prediction succeeds, that is a reason not to replicate the experiment. In other words, (6) when an experimental result is expected, a theory can be said to confirm the correctness of experimental procedure and execution even when the outcome of the experiment otherwise serves to confirm the theory.

Finally, and at a more global level, there is the strategy of (7) using the experimental result—assuming that there are no outstanding claims against its reliability—as a stepping stone to additional experimental elaborations (guided by available theory) of the initial experimental result. In other words, the experimental result justifies (in terms of the efficient use of experimental resources) the *pursuit* of such experimental exploration designed to expand on and apply the initial experimental result. Insofar as such experimental elaborations are themselves successful and fruitful the initial experimental result will have been confirmed as having been successfully conducted, and thus *if* duplicated would have yielded the same result.

In order to generate specific instances of such strategies in action, and as well to further explicate these strategies, we focused in an earlier publication on a particular class of experiments, namely, what are known as "null" experiments where the experimental effort was to develop increasingly more accurate determinations of the uncertainty involved in the determination of a null or zero result. This focus was also motivated by the conspicuous absence in the social sciences of such an obsession with "measuring nothing repeatedly"—the title of our book (Franklin and Laymon 2019).

Here we again focus on a particular sort of experiment, but this time on those where historically there was no felt need for duplication and where moreover the experimental result, in accordance with strategy (7), was seen as a stepping stone and justification of the pursuit of experimental exploration designed to expand and apply the initial experimental result. In short, these are cases where *once was enough* or depending on certain complicating historical exigencies *once could have been enough* to justify such pursuit.[1]

[1]Strict enforcement of the replication Gold Standard would suggest to the contrary that *once is not enough* and in this, as indicated by the title of her 1973 novel *Once Is Not Enough*, Jacqueline Susann would apparently agree. The implication in both the novel and the movie based on the novel was that this was a frequency reference, as in once is not enough, five times might be. In fact, this was not a frequency reference, but a time reference. Franklin discovered this when he watched the last scene in the movie. Two former lovers were discussing their relationship. He remarked, "I loved you once," as in previously. She responded, "Once is not enough," implying that he had to love her in the present.

In Chap. 2 we will examine the Meselson-Stahl experiment which decided the issue between two models of DNA replication. The experimental result was held not to require replication. A second experiment of a rather different nature was used to eliminate a third possibility, and whereby contrast that experiment was further refined so as to yield a more accurate result. These experiments set the stage for the creation of new competing accounts of DNA replication all of which—given the framework established by the Meselson-Stahl experiments—were eliminated by experimental tests leaving only the Watson-Crick model in play.

Chapter 3 deals with the discovery of the positron, the antiparticle of the electron. The positron was first discovered and reported by Carl Anderson and soon thereafter by Patrick Blackett and Giuseppe Occhialini. Neither of these experimental investigations were motivated by Paul Dirac's earlier prediction of the positron, but rather by more straightforward concerns about the absorption and production of cosmic ray particles. Thus, Anderson was quite explicit that his discovery was entirely "accidental," and while Blackett and Occhialini did mention Dirac's theory in their published report it was only after they had firmly established the existence of the positron (without any need for, or reference to, Dirac's theory) and then only to make use of that theory to support Blackett and Occhialini's proposed explanation of the production of positrons as a byproduct of a pairwise creation of positrons and electrons. Thus, with respect to the experiments of Anderson, and Blackett and Occhialini, once was enough to establish the existence of the positron and to justify in particular the pursuit of gamma ray experiments in an attempt to understand the production of positrons.

In the discovery of the omega minus hyperon, Chap. 4, the role of theory was rather different because it drove and justified the experimental effort to verify the existence of the predicted omega minus particle. The Eightfold Way, proposed by Murray Gell-Mann and Yuval Ne'eman, organized elementary particles on the basis of highly abstract considerations into certain group structures. The theory was not only highly abstract but was also what's known in particle physics as a "phenomenological theory," that is, a descriptive theory without an underlying dynamics. It had some empirical support insofar as many of the locations in the groups were filled by existing particles. But like Dirac's theory it was not generally accepted. There was, however, the opportunity, as recognized by Gell-Mann and Ne'eman, for a stunning, because not otherwise expected, predictive success in the form of the omega minus particle. Once was enough for the experimental discovery to establish the existence of the particle to the degree that further experimental and theoretical development was clearly justified, which ultimately led to the quark model of elementary particles.

At the time of Mendel's experiments on plant hybridization in pea plants (Chap. 5), there was no existing theory of hybridization. It is unclear whether Mendel had his reproductive and inheritance model, which included dominant and recessive characters, in mind when he began his work, or whether it was suggested to him by his early results. His experiments on pea plants not only suggested the laws of segregation and independent assortment, which are the basis of modern genetics, but also provided strong evidence for those laws. The former law states that variation for contrasting traits is associated with a pair of factors which segregate to individual reproductive cells. The latter states that two or more of these factor-pairs assort independently to individual reproductive cells. The experiments also provided evidence that those factors, which we now associate on a deeper biological level with genes, come in two types, dominant and recessive. And while Mendel's results once they had been rediscovered were in some respects experimentally duplicated, those duplications were of only limited importance since by far the dominant effort was to expand on Mendel's results by extending their application and tying Mendel's factors more specifically to their biological underpinnings. In short, once was very much enough to get a major experimental and theoretical endeavor under way.

In Chaps. 6 and 7 we will discuss two episodes which bear on the question of whether physical laws satisfy parity conservation or space-reflection symmetry. In Chap. 6 we will discuss the experiments of Richard Cox and his collaborators, as well as those of Carl Chase, Cox's student, which, at least in retrospect, showed that parity symmetry was violated. The experiments involved the double scattering of electrons from β decay. The second scattering showed a left-right (90°/270°) asymmetry. At the time, no one, including the experimenters themselves, realized the significance of these results. We will discuss whether these results did, in fact, demonstrate parity nonconservation and why they were overlooked.

Chapter 7 will discuss the discovery of parity nonconservation. In the early 1950s the physics community was faced with a vexing problem, the $\theta - \tau$ puzzle. There were two elementary particles which, on one set of criteria, mass and lifetime, seemed to be the same particle. On another set of criteria, spin and parity, they seemed to be different particles. Solutions using accepted physics failed to solve the problem. In 1956, T.D. Lee and C.N. Yang suggested that the violation of parity symmetry, a radical solution, would solve the problem. They proposed several experiments that would test their hypothesis. The first, the β decay of oriented nuclei was performed by C.S. Wu and her collaborators. They observed an asymmetry in the decay, more electrons were emitted opposite to the nuclear spin than along the spin direction, thus, demonstrating parity nonconservation. Similar results were obtained in a second set of experiments on $\pi \rightarrow \mu \rightarrow e$ decay by Garwin, Lederman, and Weinrich and by Friedman and Telegdi. The question of whether just one of these experiments would have sufficed to demonstrate parity violation will be discussed.

Chapter 8 seems to be a textbook example of experiment testing theory. In order to solve the question of why electric charge is quantized, all electric charges are integral multiples of the charge of the electron, Paul Dirac hypothesized the existence of magnetic monopoles, single North or South magnetic poles, something which had never been observed. Dirac's theory also predicted the strength of such a magnetic pole. In 1982 Blas Cabrera reported an event which fit Dirac's prediction very well. Yet Cabrera made no discovery claim nor was Dirac's theory regarded as confirmed. The reasons for this are discussed.

In our concluding chapter we will abstract from these cases what lessons can be learned regarding replication as the Gold Standard and the alternative means, as sketched out in Franklin's epistemology of experiment, to if not insure at least maximize the chance that an experiment has been well executed.

References

Franklin, A. 2016. *What makes a good experiment?*. Pittsburgh: University of Pittsburgh Press.

Franklin, A., and R. Laymon. 2019. *Measuring nothing, repeatedly*. San Rafael, CA: Morgan and Claypool.

Janz, N. 2015. Bringing the gold standard into the classroom: Replication in university teaching. *International Studies Perspectives*, 1-16.

Jasny, B.R., G. Chin, et al. 2011. Again, and again, and again…. *Science* 334: 1225.

Kepes, S., and M.A. McDaniel. 2013. How trustworthy is the scientific literature in industrial and organizational psychology? *Industrial and Organizational Psychology: Perspectives on Science and Practice* 6: 252–268.

National Academies of Sciences, E., and Medicine. 2019. *Reproducibility and replicability in science*. Washington, DC, The National Academies Press.

Norton, J. 2015. "Replicability of Experiment. *Theoria: An International Journal for Theory, History and Foundations of Science* 30: 229–248.

Part I
Once Was Enough

Chapter 2
The Meselson-Stahl Experiment: "the Most Beautiful Experiment in Biology"

2.1 Introduction

In our introductory chapter we made the point that replicating an experimental result is not a necessary condition for its acceptability. And while it may be a good idea in some cases, there is no basis for making it a general requirement for the validation of an experimental result. Rather the more reasonable general requirement is that there be evidential or theoretical support for the conditional that if the experiment were replicated then the same result would be obtained. But as the title of our essay suggests, sometimes once is enough for acceptability. In such cases, no additional validation is needed, and from a historical perspective was not called for. Once was enough.

The Meselson-Stahl DNA replication experiment was judged to be "the most beautiful experiment in biology" (J. Cairns, quoted in Holmes 2001, p. 429). So beautiful in fact that it serves as a paradigm for once being enough. The experiment was extensively reviewed and discussed before publication by many of the core researchers in the relevant research community. The response was uniformly positive with no calls for replication. Nor was the issue of replication raised after publication or in the many textbook accounts that followed. The question then is why was once was enough for this experiment?

2.2 Contenders for the Mechanism of DNA Replication

We begin with the theoretical and experimental background that provided the motivation and context for the Meselson-Stahl experiment.[1] In 1953 Francis Crick and James Watson proposed a three-dimensional structure for deoxyribonucleic acid (DNA): two polynucleotide chains helically wound about a common axis (Watson and Crick

[1]For an excellent history of this complex episode and complete references see Holmes (2001).

A. Franklin and R. Laymon, *Once Can Be Enough*,
https://doi.org/10.1007/978-3-030-62565-8_2

1958). This was the famous "Double Helix" where the chains were bound together by combinations of four nitrogen bases–adenine, thymine, cytosine, and guanine. The nitrogen base on one chain is hydrogen-bonded to the base at the same level in the other chain. Because of structural and binding requirements Watson and Crick postulated that only the base pairs adenine-thymine and cytosine-guanine were to be allowed. This explained the previously observed regularity, known as Chargaff's rules, that the amount of adenine contained in the DNA of any species was approximately equal to the amount of thymine. The same was true for cytosine and guanine. Each chain is thus complementary to the other. If there is an adenine base at a location in one chain there is a thymine base at the same location on the other chain, and vice versa. The same applies to cytosine and guanine. The order of the bases along a chain is not, however, restricted in any way and it is the precise sequence of bases that carries the genetic information—or so it was claimed.

Watson and Crick were quite aware that while the proposed Double-Helix structure left many questions unanswered it did set the stage for how those questions were to be answered especially with regard to genetic duplication.

> For the moment, the general scheme we have proposed for the reproduction of deoxyribonucleic acid must be regarded as speculative. Even if it is correct, it is clear from what we have said that much remains to be discovered before the picture of *genetic duplication* can be described in detail. What are the polypeptide precursors? What makes the chains unwind and separate? What is the precise role of the protein? Is the chromosome one long pair of deoxyribonucleic acid chains, or does it consist of patches of the acid joined together by protein? (Watson and Crick 1953, p. 966).

In short, if DNA was to play a crucial role in genetics there had to be a mechanism or at least some sort of initial performance script for the replication of the molecule. And as had become increasingly evident, DNA and its replication did indeed play a central role in genetics. But exactly how? As summarized by Meselson and Stahl:

> Studies of bacterial transformation and bacteriaphage infection strongly indicate that deoxyribonucleic acid (DNA) can carry and transmit hereditary information and can direct its own replication. Hypotheses for the mechanism of DNA replication differ in the predictions they make concerning the distribution among progeny molecules of atoms derived from parental molecules (Meselson and Stahl 1958, p. 671, footnotes omitted).

In fact, by this time there were three different mechanisms proposed for the replication of the DNA molecule, now assumed to have the Double-Helix structure proposed by Watson and Crick, which were reviewed in Delbruck and Stent (1957) and illustrated in Fig. 2.1.

The first, proposed by Gunther Stent and known as *conservative replication*, held that the pair of strands of the parent DNA replicate as an intact unit. This yields a first generation which consists of the original parent DNA molecule and one newly-synthesized DNA molecule of composition identical to that of the parent. The second generation will consist then of the parental DNA molecule and three newly-synthesized DNA molecules—all of identical composition and structure.

The second proposed path to replication, known as *semiconservative replication* is when each strand of the parental DNA acts as a template for a second newly-synthesized complementary strand, which then combines with the original strand to

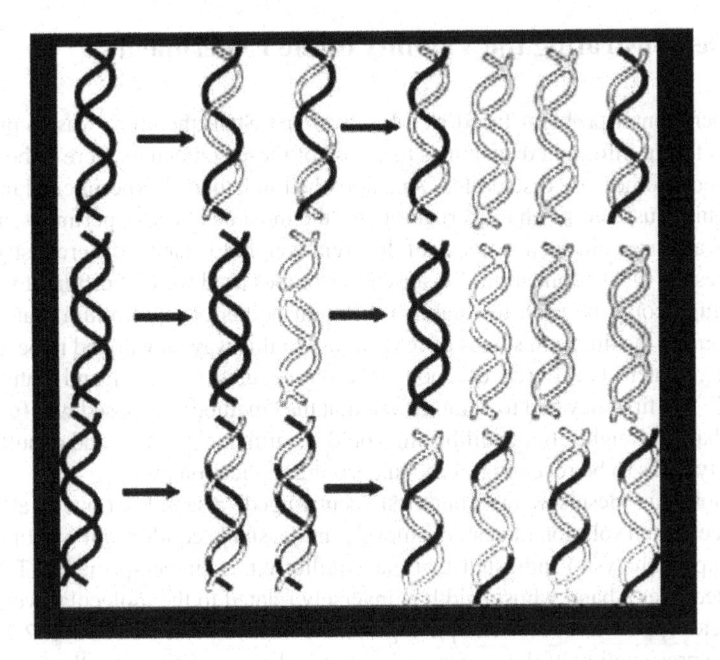

Fig. 2.1 Proposed models of DNA replication: Semi-Conservative (top); Conservative (middle); Dispersive (bottom).

form a DNA molecule. In other words, the strands somehow unwind, where each strand now self-replicates and then rejoins the original template (with compensating adjustment for polarity). Thus, the first generation consists of two hybrid molecules, each of which contains one strand of parental DNA and one newly-synthesized strand. The second generation consists of two hybrid molecules and two newly synthesized DNA molecules. This was the approach favored by Watson and Crick (Watson and Crick 1953).

In the third mechanism, proposed by Max Delbrück, and known as *dispersive replication*, the parental DNA chains break at (unspecified) intervals and the parental segments combine with new segments to form the daughter strands. This approach was proposed primarily as a way of coming to terms with the "unwinding" problem implicit in the need for the strands of DNA to somehow separate before replication.[2] It is important to note that *dispersive replication* is not necessarily incompatible with either *conservative* or *semiconservative replication*. This because it was primarily a general approach to dealing with the unwinding problem and thus its parameters could be adjusted, if needed, so as to remain compatible. As we shall see, this flexibility will play an important role when considering the confirmational scope of the Meselson and Stahl experiment.

[2]See Delbruck and Stent (1957) and Holmes (2001, pp. 15–29, 393–394).

2.3 Demonstrating the Viability of the Experiment

The experimental problem faced by Meselson and Stahl therefore was to provide the basis for an informed decision as to which of these contenders were to be either eliminated or otherwise discredited. And here the fundamental experimental innovation—which justified its characterization as "the most beautiful experiment in biology"—was to use nitrogen isotopes of different weight to "label" different stages in the process of DNA replication. Because the isotopes used were of different weight, a centrifuge could be used to establish different points of equilibrium that would correspond to the different stages of replication. In this way they hoped to be able to detect "small density differences among macromolecules (Meselson and Stahl 1958, p. 672).". But first they had to demonstrate that the "method" proposed was feasible, that is, that the sought-after equilibrium would be sufficiently stable and of sufficient sensitivity so as to be revealing of the underlying replication process.

To show this Meselson and Stahl first "centrifuged" a sample of bacterial lysate in a concentrated solution of cesium chloride until a stable equilibrium was obtained which (upon analysis) indicated that "at equilibrium a single species of DNA is distributed over a band whose width is inversely related to the molecular weight of that species (pp. 672–673)." The photographic evidence presented in Fig. 2.2 gives a striking presentation of the stages of the approach to equilibrium as the centrifuge did its work.

They next demonstrated that "[i]f several different density species of DNA are present, each will form a band at that position where the density of the CsCl solution is equal to the buoyant density of that species (p. 673)." To show this Meselson and Stahl prepared test samples that contained only N^{14} or N^{15}, two isotopes of different density. Once again they were successful as readily shown in both the photograph obtained by absorbing ultraviolet light and in the graph showing the intensity of the signal obtained with a densitometer (see Fig. 2.3). In addition, the separation between the two peaks suggested that they would be able to distinguish an intermediate band composed of hybrid DNA from the heavy and light bands.

The procedure used here to demonstrate the viability of the experimental design is a close variant of the strategies for validation of an experimental result that we discussed in the Introduction, what Franklin has referred to as an "epistemology of experimentation." As such, it played a crucial role in the determination that when it came to the Meselson and Stahl experiment once was enough.

2.4 The Meselson and Stahl Experiment and Its Result

The basic strategic design of the experiment was to "label" a sample of *E. coli* bacteria with N^{15}, a heavy nitrogen isotope, by growing it in a medium containing ammonium chloride that had been prepared with N^{15} as its nitrogen (growth nutrient) component. Once the bacteria had been allowed to replicate, the N^{15} label was changed to

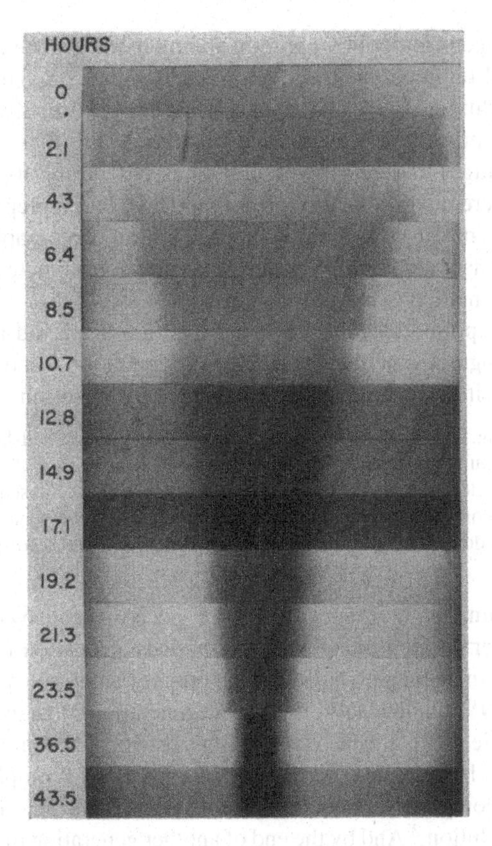

Fig. 2.2 Ultraviolet absorption photographs showing successive stages in the banding of DNA from *E. coli*. The number beside each photograph gives the time elapsed after reaching 31,410 rpm. *Source* Meselson and Stahl (1958)

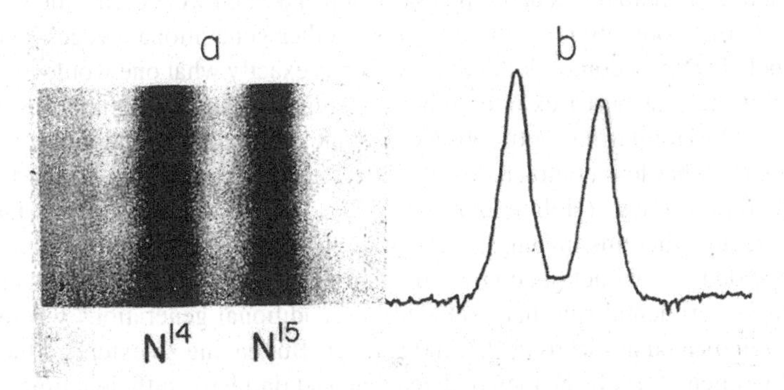

Fig. 2.3 **a** The resolution of N^{14} DNA from N^{15} DNA by density gradient centrifugation. **b** A microdensitometer tracing showing the DNA distribution in the region of the two bands in part **a**. *Source* Meselson and Stahl (1958)

N^{14} (a lighter nitrogen isotope) by abruptly adding to the growth medium a tenfold excess of $N^{14}H_4Cl$ (ammonium chloride prepared with N^{14}). Samples were taken just before the addition of N^{14} and at intervals afterward for several generations. The result of this procedure was to have samples of the *E. coli* labeled as N^{15} during its initial stages of replication and then as N^{14} during its later stages. The cell membranes were next broken to release the DNA into a prepared solution and the samples were centrifuged and ultraviolet absorption photographs taken. In addition, the photographs were scanned with a recording densitometer. See Fig. 2.4 for a schematic representation of the process.

The results as reported by Meselson and Stahl are displayed in Fig. 2.5 which includes the photographs of the centrifuge concentration bands as well as the corresponding densitometer traces. As summarized by Meselson and Stahl:

> It may be seen in Fig. [2.5] that, until one generation time has elapsed, half-labeled molecules accumulate, while fully labeled DNA is depleted. One generation time after the addition of N^{14}, these half-labeled or 'hybrid' molecules alone are observed. Subsequently only half labeled DNA and completely unlabeled DNA are found. When two generation times have elapsed after the addition of N^{14}, half-labeled and unlabeled DNA are present in equal amounts (p. 676).

Relating their summary more specifically to Fig. 2.5 we begin by noting that "[o]ne generation time after the addition of N^{14}" corresponds to location 1.0 where it can be readily seen that a single band of half-labeled (progeny) molecules has evolved from its (parental) fully N^{15} labeled DNA. It is here at generation 0 that the growth medium was abruptly modified with the introduction of an excess of N^{14} prepared ammonium chloride. And from here on one can see the emergence of "completely unlabeled" DNA—where by "completely unlabeled" Meselson and Stahl mean having evolved in the N^{15} modified solution.[3] And by the end of another generation time at 1.9 or thereabouts one can see that there now have evolved two bands and corresponding peaks that indicate the presence of half-labeled and unlabeled DNA in equal amounts.[4]

But there is more to be seen, though Meselson and Stahl were oddly silent on this. And that is that soon after 1.9 there begins another generational process whereby the unlabeled DNA becomes dominant—which is exactly what one would expect on the basis of the Watson-Crick double helix assumed to undergo semi-conservative replication. As Holmes has aptly observed "[t]he integration of the figure with the text was somewhat less complete, as the latter made reference only to the lapse of "two generation times" (Holmes 2001, p. 377). We'll come back to this later but for now we note that this incomplete integration was the result of Meselson having hastily decided to conduct his own replication of the experiment by conducting an additional experimental run that extended into additional generations where those results are indicated at locations 2.5, 3.0 and 4.1. But having so extended the range of his experimental results he then realized that he didn't have sufficient time or page

[3]Often referred in textbooks as being N^{14} labeled as opposed to unlabeled.

[4]It should be noted that the consistent locations of the various bands as a function of their association with the labeling was determined and validated by the *calibration* procedure of superimposing the data images at generation points 0 and 1.9 as shown in the penultimate data row.

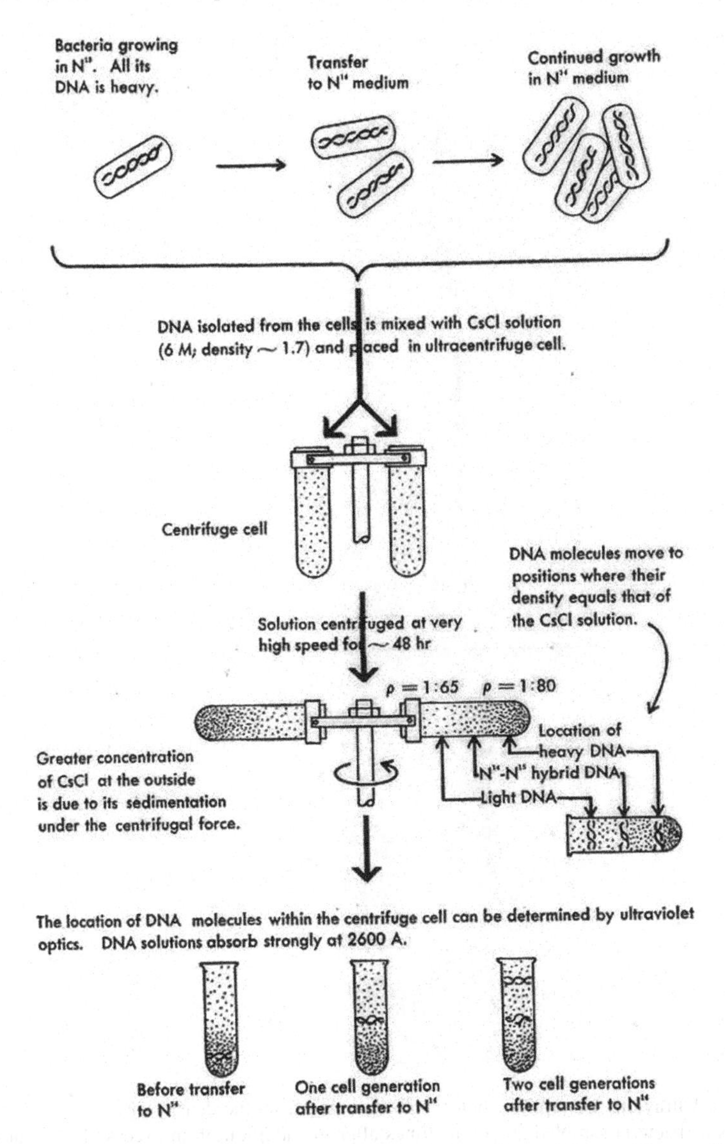

Fig. 2.4 Schematic sketch of the Meselson-Stahl experiment. *Source* Watson (Watson 1965)

space to more fully incorporate this additional data into the text. For the time and page limit constraints Meselson and Stahl were operating under see (Holmes 2001, pp. 371–375).

We have so far been less than precise in specifying what constitutes the experimental "result" which according to the "Gold Standard" is somehow to be replicated. In fact, what the Meselson and Stahl experiment reveals is that there were three separate "results" that were claimed for the experiment. First, there was what is often

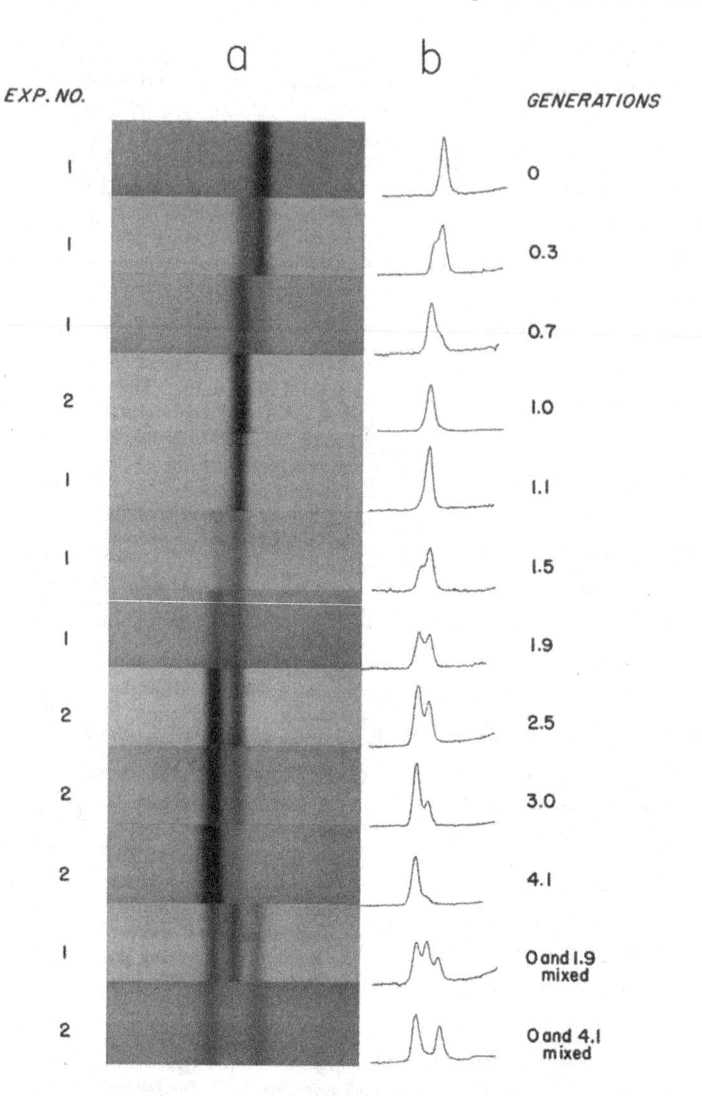

Fig. 2.5 **a** Ultraviolet absorption photographs resulting from the density-gradient centrifugation of lysates of bacteria sampled at various times after the addition of an excess of ^{14}N substrates to a ^{15}N-labeled culture.... Regions of equal density occupy the same horizontal position on each photograph. **b** The microdensitometer tracings of the adjacent photographs.... The degree of labeling of a species of DNA corresponds to the relative position of its band between the bands of fully labeled and unlabeled DNA shown in the lowermost frame, which serves as a density reference. A test of the conclusion that the DNA in the band of intermediate density is just half-labeled is provided by the frame showing the mixture of generations 0 and 1.9. *Source* Meselson and Stahl (1958)

described as the "raw data" result. Here, as we have seen, that raw data is concisely displayed in the photographs and graphs contained in Fig. 2.5 of the Meselson-Stahl report. Indeed, that display is compelling in its simplicity and economy. So, with respect to this result, the demand for replication would be a demand that this raw data (and taking into account anticipated variations) be reproduced.

But there is also a sense of "experimental result" that takes into account the theoretical relevance of the raw data. And here one might have expected Meselson and Stahl to move directly to a consideration of how their data was consistent with the semi-conservative replication of the Watson-Crick Double-Helix. Meselson and Stahl, however, did not do this. Instead, they very deliberately determined the relevance of their data for the question of whether DNA replication proceeded conservatively or semi-conservatively without regard as to whether the Watson-Crick Double-Helix was the correct description of the DNA molecule. In order to facilitate this determination, they introduced the more theory-neutral notion of a "subunit" where the experimental results would be expressed in terms of their relevance for the replication of such subunits. By way of later explaining why the decision was made to report the results of the experiment in this more theory neutral though consequently more abstract way, Meselson explained that:

> I wanted to write the paper, not the way we approached the experiment, but the other way back. What I mean by that is that we did the experiment in order to distinguish between different models for DNA replication. But I wanted to put all that aside and say, what if all you knew was what was in the experiment.. .. So, I spent a long time trying to write down sentences which would say in the [most] rigorous and briefest form, what it was that the experiment said: and finally got down to this notion that it says there is something which has two subunits, which contain heavy nitrogen, and which come apart. .. and become associated with newly synthesized sub-units. (Meselson as quoted in Holmes 2001, pp. 374–375).

Meselson and Stahl did not, however, attempt to give a formal definition of what constituted a subunit. Instead they employed the strategy of taking advantage of its ordinary meaning and then using the concept and thereby letting its technical meaning show itself. In particular, the notion of a subunit was used to express three conclusions that followed from the evidence. Looking ahead to what these conclusions have to say the reader is advised to keep Fig. 2.6 in mind. The first of these conclusions is that:

> 1. *The nitrogen of DNA is divided equally between two subunits which remain intact through many generations* (p. 676).

The argument for this conclusion consists of two parts where the first is:

> The observation that parental nitrogen is found only in half-labeled molecules at all times after the passage of one generation time demonstrates the existence in each DNA molecule of two subunits containing equal amounts of nitrogen (p. 676).

Stated more fully, the supporting experimental fact is that after the transformation of the growth medium from a N^{15} to N^{14} based ammonium chloride only half-labeled molecules were found. Assuming then that a subunit is specified by its label-identified molecular content, it follows that each DNA molecule consists of two subunits containing equal amounts of nitrogen. The second part of the argument is:

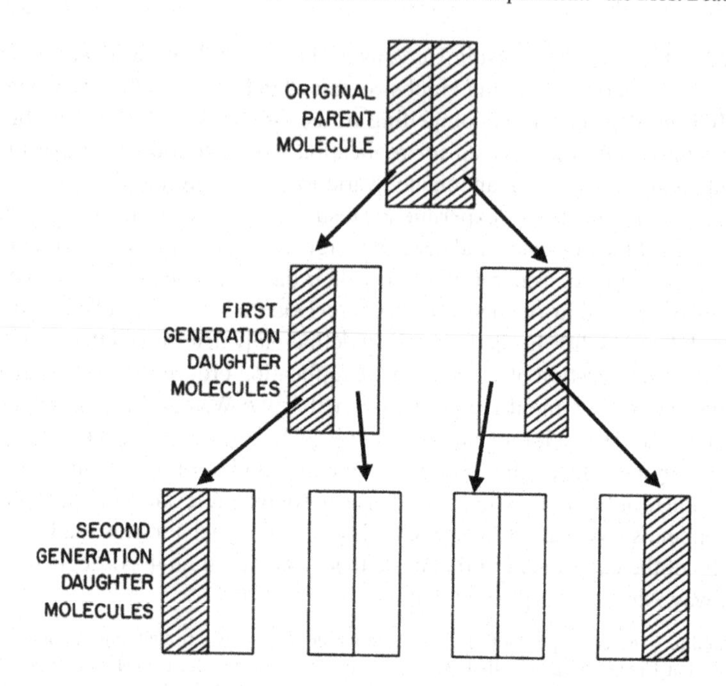

Fig. 2.6 Schematic representation of the conclusions drawn in the text from the data presented in Fig. [2.5]. The nitrogen of each DNA molecule is divided equally between two subunits. Following duplication, each daughter molecule receives one of these. The subunits are conserved through successive duplications. *Source* Meselson and Stahl (1958) with conservation paths added

> The finding that at the second generation half-labeled and unlabeled molecules are found in equal amounts shows that the number of surviving parental subunits is twice the number of parent molecules initially present. That is, the subunits are conserved (p. 676).

Exactly why this is so is not so easy to see. The experimental fact appealed to is that by the conclusion of the second generation (i.e., at 1.9 on Fig. 2.5) the number of half-labeled was equal to the number of unlabeled DNA molecules. Now how is it supposed to follow from this that "the number of surviving parental subunits is twice the number of parent molecules initially present"? Answer: Meselson and Stahl were implicitly making use of the assumption that *subunits replicated once and only once* (before going on to recombination and a repeat of the replication process). Not only was this well accepted but it was also *shared* by the conservative and semi-conservative accounts of DNA replication. With this implicit assumption in hand, the argument goes through straightforwardly.[5]

---

[5]To see how this goes trace the replication paths indicated in Fig. 2.5 and note that at the end of the first generation there are now (as per the first part of the first conclusion) twice as many subunits are there were parental subunits to begin with. And then at the next generation there will again (as per the second part of the first conclusion) be twice as many progeny but where only one half will be half labeled. Thus, the number of half-labeled subunits at the end will be twice the number of parental subunits at the initial stage of the process.

The second subunit conclusion is that:

2. *Following replication, each daughter molecule has received one parental subunit* (p. 676).

The argument for this conclusion is thankfully not so complicated.

The finding that all DNA molecules are half-labeled one generation time after the addition of N^{14} shows that each daughter molecule receives one parental subunit. If the parental subunits had segregated in any other way among the daughter molecules, there would have been found at the first generation some fully labeled and some unlabeled DNA molecules, representing those daughters which received two or no parental subunits, respectively (p. 676).

To see how this argument works consider first a parental DNA molecule at the time when the growth medium was abruptly changed from N^{15} to N^{14} based ammonium chloride. Because all such parents are composed of two N^{15} subunits two replications will occur. The experimental evidence shows that the offspring consist of exclusively half-labeled molecules. Here we remind the reader that once the growth medium was changed from N^{15} to N^{14} this means that the molecules produced by replication at this stage will be "unlabeled." The only way to obtain the experimental result of all offspring being half-labeled is for each of the offspring to have received an original (parental) unreplicated N^{14} subunit.

On the basis of the above conclusion, Meselson and Stahl present the following corollary.

3. *The replicative act results in a molecular doubling* (p. 676).

And since the reasoning here is straightforward we'll present it without comment.

This statement is a corollary of conclusions 1 and 2 above, according to which each parent molecule passes on two subunits to progeny molecules and each progeny molecule receives just one parental subunit. It follows that each single molecular reproductive act results in a doubling of the number of molecules entering into that act (pp. 676–677).

What is the bottom line? Why all this effort on the part of Meselson and Stahl? Ordinarily when considering the application of an experimental result, there is initially a theoretical prediction which is then tested by the experiment. Thus, if the prediction is not substantiated by the experimental result, then (assuming other complications don't intercede) the theory is said to have been disconfirmed. And if the prediction is found to be true, then the theory is confirmed but only in the sense of having been found to be consistent with the experimental data. But Meselson and Stahl wanted more than this. What they wanted to show was that they could *deductively derive* semi-conservative replication from the experimental data (conjoined of course with low level and uncontroversial theoretical assumptions). Thus, semi-conservative replication was not merely consistent with the data but it was a *deductive consequence* of the data.[6] This then is the *second result* of the Meselson and Stahl experiment.

[6] In a later chapter on the discovery of the positron we'll review another instance where the theoretical results were deduced from the experimental data (augmented by non-controversial theoretical assumptions) when Patrick Blackett made such a derivation in order to clearly demarcate what his experimental results demonstrated from their role in confirming Dirac's hole theory of the production and nature of the positron.

Given that semi-conservative replication had been experimentally demonstrated to be the case with respect to "subunits," the next task was to apply that result so to show that the Watson-Crick DNA replicated in a semi-conservative fashion—which was what Watson and Crick had proposed for their DNA model. But showing this was *immediate* given the evident (though not expressly stated) assumption that each of the double-helix chains was a *particular embodiment* of a Meselson and Stahl *subunit*. Thus, Meselson and Stahl were able to simply state their *third result* without further ado:

> The results of the present experiment are in exact accord with the expectations of the Watson-Crick model for DNA duplication (p. 678).

There was, however, a caveat issued to the effect that it was not thereby demonstrated that the Watson-Crick model for DNA duplication was the only account that would satisfy the requirement that subunits replicate in a semi-conservative fashion.

> However, it must be emphasized that it has not been shown that the molecular subunits found in the present experiment are single polynucleotide chains or even that the DNA molecules studied here correspond to single DNA molecules possessing the structure proposed by Watson and Crick (p. 678).

And in this Meselson and Stahl were prophetic since soon thereafter two such alternative accounts were proposed that were consistent with semiconservative replication. The first was that the DNA molecule consisted of two double helix molecules vertically bound together end to end that then separated, individually replicated and then vertically bonded with its progeny. This was disposed of relatively quickly by the experimental work of Meselson's student Ronald Rolfe. See (Holmes 2001, p. 392). Liebe Cavalieri also favored a dimer as the DNA structure, but this time one composed of two laterally bonded double helix molecules. On this approach, the replication process also begins with a separation of the double helix units, where each (now separated) double helix replicates, and then bonds with its progeny to form a new DNA molecule. The original parental double helix molecules are thus "conserved" after replication but are now bonded with a different partner. Cavalieri, however, clearly states that "[r]eplication of the parental molecule is therefore *semiconservative*, but replication of the double helix is *conservative*" (Cavalieri and Rosenberg 1961, pp. 343–344, emphasis added)." See also: "The experiment of Meselson & Stahl showed clearly that *E. coli* DNA replicates semiconservatively (Cavalieri and Rosenberg 1962, p. 260). Thus, Cavalieri was careful to preserve consistency with the Meselson and Stahl subunit conclusions. In other words, Meselson and Stahl's subunit is now identified by Cavalieri with an entire double helix molecule and not just one of the chains. Achieving a definitive experimental determination on Cavalieri's proposal took a while but eventually was decided by John Cairns in favor of a single Watson-Crick double-helix as the structure of the DNA molecule (Cairns 1962; Cairns 1963).

2.5 Once Was Enough

Even before the Meselson and Stahl paper was published, the results were known to many of those working in the field by means of letters, phone calls, seminars, and visits.[7] The reception was definitely positive. So, for example, and prior to the publication of the results, Christian Anfisen, who was preparing his text *The Molecular Basis of Evolution*, requested photographs of the results for inclusion. Insofar as textbooks are evidence of what is or was accepted, Anfisen's request is a positive indication that the Meselson-Stahl results were well received.

Similarly, for when Meselson visited Stent at Berkeley. When Stent saw the photographs of the centrifuge results, he realized that DNA did not replicate conservatively as he had proposed, and complained cheerfully about the "devilish, hellish experiment" (Holmes 2001, p. 324), whose results were not open to challenge. At a seminar given by Meselson during his visit, Stent introduced him as "The Mozart of Molecular Biology" (Holmes 2001, p. 325). Still, Stent had some reservations. In a letter to Delbrück he wrote, "Things *do* look good for the WC [Watson-Crick] mechanism, I must admit (very generous of me, isn't it?), but I *still* can't see how the intimate connection between replication and genetic exchange can be accounted for from this point of view" (quoted in Holmes 2001, p. 325)." What Stent had in mind here was, for example, what he described in the same letter as "'information' transmission during phage reproduction. Thus, even assuming the correctness of the Meselson and Stahl results the problem was to determine the connection between DNA replication and the more general reproductive process. Delbrück was similarly positive. In a letter to a friend he wrote, "Meselson is making earth-shaking discoveries in the replication of DNA; and every afternoon he and Sinsheimer and I have been having endless discussions (with tea) about this (quoted in Holmes 2001, p. 325)."

Meselson sent a mimeographed copy of his paper to Maurice Wilkins, who would later share the Nobel Prize with Watson and Crick for the discovery of the structure of DNA. Wilkins replied, "Thank you very much for your letter and the mss [manuscript] describing your elegant and definitive experiments.. .. as a result of your experiments I personally begin to feel real confidence in the Watson and Crick duplication hypothesis (quoted in Holmes 2001, p. 387)."

While such evidence is admittedly limited in scope, the personalities involved give it significant impact. Thus, for the leading figures involved, the Meselson-Stahl experiment was seen to have had provided strong evidence for semiconservative replication. This conclusion is further fortified by the textbook accounts that came not long after. Watson's 1965 text, *Molecular Biology of the Gene* (1965), contains a section titled, "Solid Evidence in Favor of DNA Strand Separation," in which he describes the Meselson-Stahl experiment (Fig. 2.4 is from Watson's book), although oddly he makes no mention of either Meselson or Stahl. Typical of texts written later in the 1970s are Albert Lehninger's *Biochemistry* (Lehninger 1975) and Lubert Stryer's *Biochemistry* (Stryer 1975). Lehninger judged the experiment to be a "crucial

[7] Younger readers may need to be reminded that email was not always available.

experiment" because it "[i]ngeniously [and] conclusively proved that in intact living *E. coli* cells DNA is replicated in the semiconservative manner postulated by Watson and Crick (Lehninger 1975, p. 892)." Stryer in a section titled "DNA Replication is Semiconservative" refers to the Meselson-Stahl experiment as a "critical test." (Stryer 1975, p. 568). He does, however, cite the caveat of the Meselson-Stahl paper discussed earlier, that other structural accounts were still possible.

So much for the historical fact that once was deemed enough. Our question now is what is it about the Meselson and Stahl experiment that *justifies* the historically rendered judgment?

First, Meselson and Stahl carefully demonstrated that the experimental apparatus process had the requisite sensitivity and reliability to detect the differing separation effects predicted by the conservative and semi-conservative accounts of DNA replication.

Second, because quantitative accuracy was not at a premium, the succession of developing bands, presented photographically (as well as in more traditional graphical form), could readily be seen, even by the naked eye, as indicative of the separation of the differently labeled DNA and progeny. This stands in contrast for example to null experiments where "replication" is called for in the form of increasingly more accurate and inclusive determinations of the sought-after null value.[8]

Third, no worrisome candidates were proposed for confounding causes. At least none that anyone could think of. And certainly, none that seemed capable of rendering the results as some sort of accidental coincidence that would not be repeated in a replication designed to deal with such a contingency.[9]

Fourth, Meselson himself recognized that it was "important to show that the experiment was repeatable" (Holmes, 2001, p. 347). But insofar as Meselson repeated the experiment it was, as discussed above, an "improved" replication in the sense that the experiment was continued over several additional DNA replication cycles all with the functionally equivalent result of subunit conservation. Thus, Meselson's replication was designed not just to repeat (and in fact it did not fully do so) but served to "extend" the experiment from two to "four generations of bacterial duplications."

Fifth, insofar as the semiconservative Watson-Crick account was seen, even before the Meselson and Stahl experiment, as more likely than Stent's conservative account, the predictive success of Watson-Crick counted as reason not to replicate the Meselson and Stahl experiment. We hasten to add, however, that in this case we

---

[8]For an extensive analysis of null experiments and the sense in which they satisfy demands for replication see Franklin and Laymon (2019).

[9]There is, though, a caveat that we need to make. There emerged on the basis of further experimentation and theoretical development a realization that the sharpness of the separation bands reported by Meselson and Stahl was due in part to an unrecognized but uniformly consistent fragmentation of the DNA molecule that occurred during the preparation of the DNA samples. But the defect was not claimed in any way to lend experimental support for conservative replication, and moreover new experiments continued to support semi-conservative replication. For the details see Holmes (2001, pp. 395–397). As Thomas Kuhn would have described it, these further developments were part of *normal science* conducted in response to the Meselson and Stahl experiment understood as a *paradigm* (Kuhn 1962, pp. 23–42).

don't want to put all that much weight on this consideration. We will, however, give a more convincing example in a later chapter on the discovery of the Omega minus particle.

2.6 Dealing with Dispersive Replication

The reader will have noted that Meselson and Stahl have so far been silent on the relevance of their results for the dispersive replication account. And while they were also silent on that relevance for the conservative account, that silence is readily explained by the fact that the conservative account was so *clearly* inconsistent with the experimental results and thus this inconsistency did not need to be expressly stated. The situation, however, with respect to dispersive replication is rather more complicated. The problem was that the raw data as incorporated in the Meselson and Stahl "subunit" conclusions do *not* refute dispersive replication. Knowing this, Meselson and Stahl conducted a *different set experiments* in order to dispense with it. Now you might think on the basis of Meselson and Stahl's diagrammatic representation of the subunit conclusions (as represented in Fig. 2.6) that dispersive replication was eliminated from the running.[10] But that diagram is misleading because it suggests that the components of the subunits are contiguous in their location on the parts (whatever they may be) of the DNA molecule. So, if, for example, the DNA molecule is a Watson-Crick double helix then a subunit would appear to exclusively occupy the entirety of a corresponding chain of the double-helix. But the data do *not* so constrain the location of the components of the subunits. This can be readily seen and understood in terms of our Fig. 2.7. Here the rectangular boxes are now to be understood as corresponding to the Watson-Crick DNA chains so that what is displayed is the possible non-contiguous location of the subunits in a way that is consistent with semi-conservative replication. As indicated, at each generational stage the subunits (while dispersed) are conserved (within the confines of the newly created DNA molecule) where in addition each subunit separates from its partner before replication.[11]

---

[10]There is an argument to be made even if it's true that dispersive replication is not formally inconsistent with semi-conservative replication, the constraints imposed by semi-conservative replication on dispersive replication are so severe and restrictive that dispersive replication is *effectively disconfirmed* because of the evident difficulties that would be involved in creating a developed and specifically testable version of dispersive replication. Alternatively stated, once was enough for the Meselson and Stahl experiment to have effectively discouraged any further *pursuit* of dispersive replication as a promising approach to understanding the replication of DNA. Viewing dispersive replication as having been effectively disconfirmed in this way provides a charitable explanation why many of the textbooks claim that the Meselson and Stahl experiment disconfirmed not only conservative replication but also dispersive replication. See, for example (Lehninger 1975, pp. 659–61) where it is claimed that the results of the Meselson and Stahl experiment "are exactly those expected from the hypothesis of semiconservative replication proposed by Watson and Crick; whereas they are not consistent with the alternative hypotheses of conservative or dispersive replication.".

[11]One can make the representation more realistic (in the sense of representing dispersion at multiple locations) by simply stacking the rectangular modules and making them of different height (and

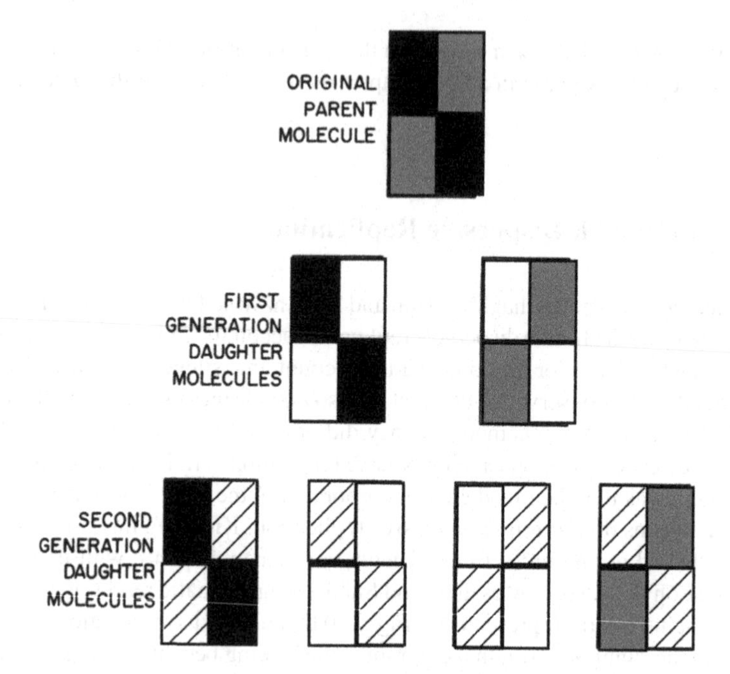

Fig. 2.7 Schematic representation of subunit replication that satisfies the requirements of both semi-conservative and dispersive replication. Adapted by the authors from Meselson and Stahl (1958)

Still, it must be kept in mind that if semi-conservative replication is correct, as the Meselson and Stahl data clearly demonstrate, then that result *constrains* dispersive replication insofar as the *reassembly* of sub-components after an initial disassociation must exactly proceed so as to preserve the initial subunit identities. Otherwise semi-conservative replication would be violated. To add some specificity to the problem, assume that the DNA molecule is in fact the Watson-Crick double helix, and that dispersive replication is correct. If so, then each of the helix chains will become *fragmented* as part of the replication process. So, lots of fragmented pieces will be floating around. What then is the mechanism whereby these now separated pieces find their way back to their original configuration and location?

Delbrück, for his part, was evidently aware of the difficulties involved in making dispersive replication consistent with semi-conservative replication. In a letter to James Taylor, Delbrück noted:

> It seems of course pretty extravagant to assume that the breaks occur at exactly reproducible points at each replication, but, who knows, there may be landmarks along the way to keep things orderly (quoted in Holmes 2001, p. 278).

internal proportionality) to accommodate different dispersion locations. In which case, the argument for consistency with semi-conservative replication carries through as before.

Similarly, as evidenced by an interview conducted by Fredric Holmes, Meselson and Stahl were also aware that while conservative replication did not refute dispersive replication it would impose a serious constraint on its application.

> Stahl's 1992 recollection that "we actually considered the far-out possibility that the break and join in each generation were precisely the same on generation to the next," resonates also with the interchanges that must have taken place between them and Delbrück, for this was the same possibility that Delbrück had raised in his letter to Taylor (Holmes 2001, p. 342).

Given then that Meselson and Stahl knew all this, the question arises: Why didn't they say so in their 1958 paper? The answer is that there were both page limits and serious time constraints on when their article had to be completed. On this see again (Holmes 2001, 371–375). So rather than go through the complexities that we've just discussed they went directly to report on a *different set of experiments* that more or less straightforwardly refuted dispersive replication.

> By way of background for these experiments, the following was known.
>
> Exposure to elevated temperatures is known to bring about an abrupt collapse of the relatively rigid and extended native DNA molecule and to make available for acid-base titration a large fraction of the functional groups presumed to be blocked by hydrogen-bond formation in the native structure (Meselson and Stahl 1958, p. 679).

As shown in Fig. 2.8a (which deals with the effects of heating N^{15} *E. coli* DNA), such collapse was indicated by the increased *density* of what remained after heating, as well as by the fact that the *molecular weight* was "reduced to approximately half that of the unheated material" (681).[12] What Meselson and Stahl then did was to perform a similar experiment but this time using hybrid *E. coli* DNA. Not surprisingly after heating, as shown in Fig. 2.8b, two concentration-density peaks resulted which "suggests that heating the hybrid molecule brings about the dissociation of the N^{15}-containing subunit from the N^{14} subunit" (681).

In order to test this possibility, Meselson and Stahl performed another density-gradient examination but this time "of a mixture of heated N^{15} DNA and heated N^{14} DNA" where the results are indicated in Fig. 2.8c. Putting this all together they concluded:

> The close resemblance between the products of heating hybrid DNA (Fig. 2.8b) and the mixture of-products obtained from heating N^{14} and N^{15} DNA separately [Fig. 2.8c] leads to the conclusion that the two molecular subunits have indeed dissociated upon heating (p. 681).

And now, at last, we get to the desired conclusion, namely, the *refutation* of dispersive replication.

> Since the apparent molecular weight of the subunits so obtained is found to be close to half that of the intact molecule, it may be further concluded that the subunits of the DNA molecule which are conserved at duplication are single, continuous structures. *The scheme for DNA duplication proposed by Delbrück is thereby ruled out* (p. 681, footnote omitted, emphasis added).

---

[12]Meselson and Stahl were here relying on a method of density and weight determination that they had developed along with Jerome Vinograd which was reported in Meselson et al. (1957).

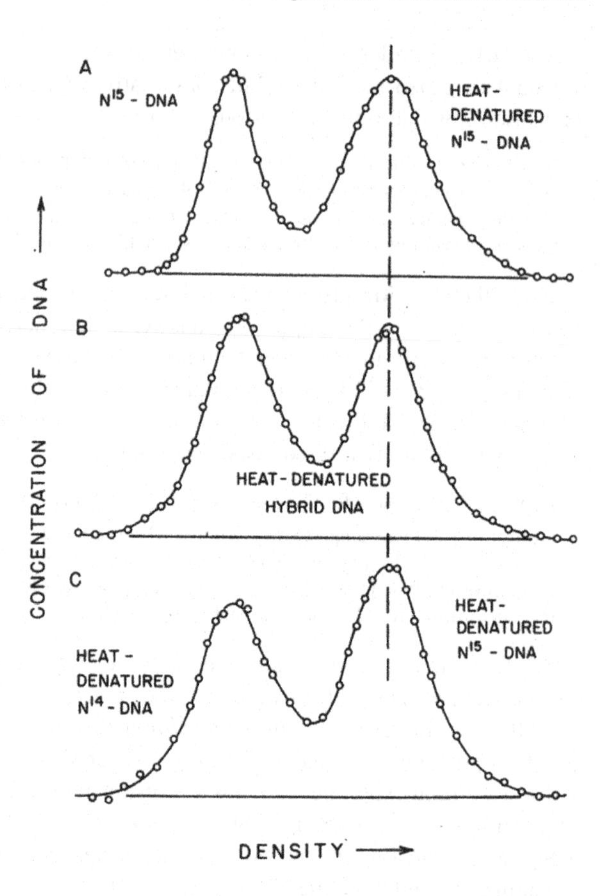

Fig. 2.8 The dissociation of the subunits of *E. coli* DNA upon heat denaturation. Each smooth curve connects points taken after 20 h of centrifugation in CSCl solution at 44,770 rpm.… **a** A mixture of heated and unheated N^{15} bacterial lysates. Heated lysate alone gives one band at the position indicated.… **b** Heated lysate of ^{15}N bacterial grown for one generation in N^{14} growth medium. Before heat denaturation, the hybrid DNA contained in this lysate forms only one band. **c** A mixture of heated N^{14} and heated N^{15} bacterial lysates. *Source* Meselson and Stahl (1958)

Note first of all, that the one-half weight result substantiates the suggestion that the subunits survive heating essentially intact. This is because if they had fragmented the experimentally determined molecular weight would have been significantly less than one-half that of the intact molecule before heating.

Second, in order to get to the conclusion that dispersive replication is ruled out Meselson and Stahl are implicitly assuming that if dispersive replication were correct then the heating of the *E. coli* DNA would have caused *additional disassociation and fragmentation* along the separation points postulated by dispersive replication. This in turn would result in a corresponding increase in density and, more importantly, a decrease in molecule weight to a value of less than the experimentally determined value of one half of the molecular weight of the unheated and intact DNA molecule.

We suppose that a diehard supporter of dispersive replication could fight back and claim, for example, that the additional disassociation was a not necessary part of dispersive replication, but by now the complexity and ad hoc nature of any salvage operation were such that no one took up the cause. There were, however, reservations expressed about the accuracy of the methods used to make the density and weight determinations. And, in fact, Meselson and Stahl had themselves noted that there were "several possible sources of error" in those determinations (Messelson and Stahl 1958, p. 673, p. 682 n12). Consequently, better methods were soon developed.[13] But for current purposes what's important is that these concerns with the accuracy and reliability of the density and weight determinations were *not* recruited in an effort to somehow resuscitate dispersive replication. In this respect, once was deemed good enough for the additional Meselson and Stahl experiments to have demonstrated that "[t]he scheme for DNA duplication proposed by Delbrück is thereby ruled out" *in the sense that* there were no calls for the replication of this *result* by either a repeat of the Meselson and Stahl experiment or by other experimental means.

2.7 Summary and Conclusions

There were three separate "results" that were claimed for the Meselson and Stahl experiment. First, there was what we have described as the "raw data" result—which in this case was concisely reported and displayed in the photographs and graphs contained in Fig. 2.5 of the Meselson-Stahl report. To describe such data as "raw" means only that the theoretical presuppositions involved were not at issue. So, for example, in this case these presuppositions dealt with the operation of the centrifuge, the measurement instrumentation and the underlying chemical processes that made labeling with different isotopes possible.

The subunit conclusions constitute a second, higher level theoretical result. Here the idea was to use the raw data as premises in an argument that would deductively entail the replication properties of the subunits. Though once again some theoretical presuppositions were required such as that the subunits replicated once and only once in each generation time.

Finally, at the highest and most specific theoretical level, there was the result of having confirmed the Watson-Crick account, i.e., that if the DNA molecule consisted of a single Watson-Crick double helix then its replication was semi-conservative and not conservative.

It is tempting to dismiss the importance of the subunits result given that the requirement of semi-conservative replication was (as per the third result) satisfied by the Watson-Crick account. Couple this with the ultimate triumph of the Watson-Crick account and the subunits result seems no more than a historical diversion. But there was benefit because the subunits result clearly set the standard that any competitor to the Watson-Crick account would have to satisfy. Thus, it opened the door for the proposal that instead of a single double-helix what was involved was a pair of attached (either end-to-end or laterally) double-helixes where each of the component helixes was conserved during replication but that the unit as a whole replicated in the semi-conservative manner required by the subunits result. In addition, reflection on the subunits result evidently led Meselson and Stahl to realize that while that result placed a significant constraint on dispersive replication it did not close the door on

[13]For a brief review of this development see Holmes (2001, 394–395).

dispersive accounts where the subunits were distributed non-contiguously over the chains of a double helix. To so close the door required, as they realized, a separate line of experimentation designed to show that the subunits had to be contiguously located on the chains of the double-helix.

For the reasons examined earlier, once was enough for the Meselson and Stahl experiment with respect to each of the three results we have identified. There were no calls for the raw data to be reproduced by means of the duplication of the experiment. Nor were there calls for a replication, possibly by means of a very different sort of experiment, of the result of having confirmed the subunit conclusion, or of having confirmed the semi-conservative result for the double helix. If on the other hand there had been good reason to replicate the confirmation of the semi-conservative result for the double helix by means of a different sort of experimental arrangement, and if that replication of result had been successful, then that success would count as evidence that *if* the Meselson and Stahl experiment were to be duplicated then essentially the same raw data would result. Moreover, such replication by means of a different sort of experiment would likely be more efficacious than a simple repetition of the original experiment because using a different sort of experiment would more likely reveal the presence of unrecognized confounding causes.

With respect to the additional experiments employed by Meselson and Stahl to refute dispersive replication, once was good enough *in the sense that* there were no calls for the replication of their claimed *result* of refutation by either a repeat of the Meselson and Stahl weight experiments or by other experimental means. Though as explained above, in recognition of certain defects in the method used by Meselson and Stahl to determined molecular weight there were—because of its importance in other experimental efforts—efforts initiated to improve the means used to determine that weight.

References

Cairns, J. 1962. Proof that the replication of DNA involves the separation of the strands. *Nature* 194: 1274.

Cairns, J. 1963. The bacterial chromosome and its manner of replication as seen by autoradology. *Journal of Molecular Biology* 6: 298–213.

Cavalieri, L., and B.H. Rosenberg. 1961. The replication of DNA: III. Changes in the number of strands in E. coli DNA during its reproduction. *Biophysical Journal* 1: 337–351.

Cavalieri, L., and B.H. Rosenberg. 1962. Nucleic acids: Molecular biology of DNA. *Annual Review of Biochemistry* 31 (258): 247–270.

Delbruck, M., and G.S. Stent. 1957. On the mechanism of DNA replication. In *The chemical basis of heredity*, ed. W.D. McElroy and B. Glass. Baltimore: Johns Hopkins University Press.

Franklin, A., and R. Laymon. 2019. *Measuring nothing, repeatedly*. San Rafael, CA: Morgan and Claypool.

Holmes, F.L. 2001. *Meselson, Stahl, and the replication of DNA: A history of "The Most Beautiful Experiment in Biology"*. New Haven: Yale University Press.

Kuhn, T. 1962. *The structure of scientific revolutions*. Chicago: University of Chicago Press.

Lehninger, A.L. 1975. *Biochemistry*. New York: Worth.

Meselson, M., and F.W. Stahl. 1958. The replication of DNA in Escherichia coli. *Proceedings of the National Academy of Sciences (USA)* 44: 671–682.

Meselson, M., F.W. Stahl, et al. 1957. Equilibrium sedimentation of macromolecules in density gradients. *Proceedings of the National Academy of Sciences* 43 (7): 581–588.

Stryer, L. 1975. *Biochemistry*. New York: W.Y. Freeman.

Watson, J.D. 1965. *Molecular biology of the gene*. New York: W.A. Benjamin.

Watson, J.D., and F.H.C. Crick. 1953. Genetical implications of the structure of deoxyribonucleic acid. *Nature* 171: 964–967.

Watson, J.D., and F.H.C. Crick. 1958. A structure for deoxyribonucleic acid. *Nature* 171: 737–738.

Chapter 3
The Discovery of the Positron

3.1 Introduction

In this chapter and the next we will discuss two experimental results from the history of particle physics that respectively established the existence of the positron and the omega minus hyperon. In each episode, a single observation of the particle in question was—at least at first glance—sufficient to persuade the physics community of its existence. These observations which were captured in photographic form are what Peter Galison has called "golden events" where there's a "single picture of such clarity and distinctness that it commands acceptance" (Galison 1997, p. 22).

An overstatement to be sure, but still correct to the extent that these discoveries were encapsulated in striking photographs of the crucial observational events which made the discoveries easy to see and understand—assuming, of course, that the viewer was armed with the necessary theoretical prerequisites. To which must be added evidence, statistical and otherwise, indicating that the photographs were to be taken at face value.[1]

We have already seen such an "golden event" in the form of Meselson's photographic portrayal of the centrifugal separation of the sequential stages of DNA replication. In short, all you needed to know could be seen at a glance, and where Meselson was scrupulous in isolating the relatively minimal theoretical underpinning needed to do so. Thus, acceptance was commanded in the sense of having been greatly facilitated. Bluntly stated, in such cases there is less of the contentious ruckus about acceptance that one would have expected for such consequential discoveries. How

[1]Galison's claim about such Golden Events encapsulated in compelling pictures is part of a more general thesis wherein he claims a distinction between two different traditions and methods of presenting experimental results. First, an "image" tradition where results are displayed pictorially. And second, where numerical results are presented in lists, tables and the like. Galison further claims a difference in the "logic" used to substantiate and apply the data so presented. Kent Staley has ably argued against any sharp distinction in the "logic" employed (Staley 1999). We agree, as further substantiated by the case studies we present here.

A. Franklin and R. Laymon, *Once Can Be Enough*,
https://doi.org/10.1007/978-3-030-62565-8_3

much less is what we intend to clarify with regard to the discovery of the positron and the omega minus hyperon.

What spurs our interest here in these golden events is the expectation that in such cases any felt need for replication, either exactly or nearly so, will be at a minimum or effectively zero. Thus, at bottom our interest is with the question of to what extent was "once enough" to command acceptance, and if so, acceptance for what ends?

We will begin with the discovery of the positron, the antiparticle of the electron. If one examines the discussions in contemporary physics texts this episode strongly resembles the discovery of the omega minus hyperon, to be discussed in the next chapter. In short, or so the story goes, there is a theoretical prediction of such a particle which is confirmed by an observation captured in a single photograph. In the case of the positron the "golden" photograph as captured by Carl Anderson is here reproduced as Fig. 3.1. As compared with its photographic contemporaries this is indeed a photograph of "great clarity and distinctness" which shows a single ionization path apparently entering from below and then getting increasingly bent once it made its way through the intervening lead plate. It certainly was "golden"

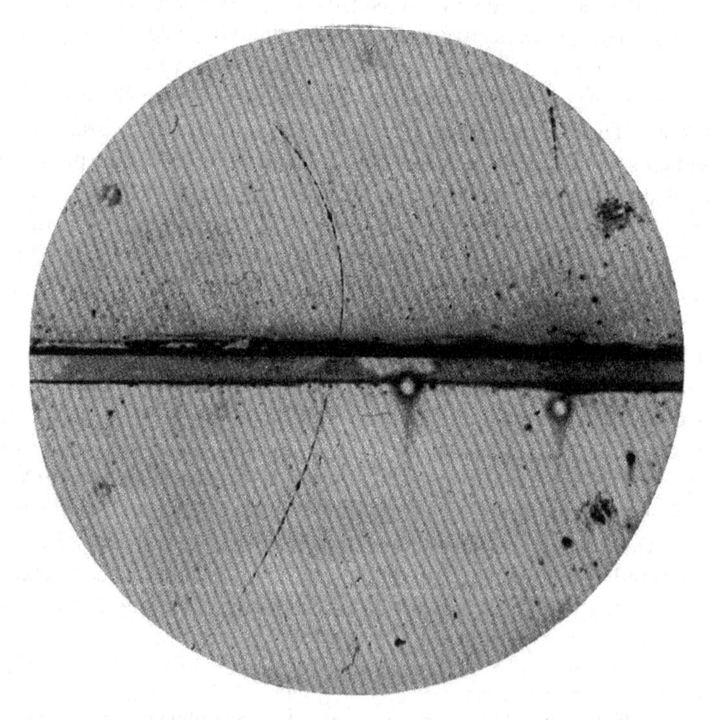

Fig. 3.1 A 63 million volt positron ($H\rho = 2.1 \times 10^5$ gauss-cm) passing through a 6 mm lead plate and emerging as a 23 million volt positron ($H\rho = 7.5 \times 10^4$ gauss-cm). The length of this latter path is at least ten times greater than the possible length of a proton path of this curvature. *Source* Anderson (1933a)

for Anderson for whom the photograph "just said, 'It's got to be a positive electron' (Anderson 1966, p. 22)."

The physics textbooks tend to treat such episodes as a simple convergence of experimental discovery and worthy theoretical prediction waiting to be confirmed. As retold in a well-known physics textbook:

> Anderson was studying cosmic-ray particles in his cloud chamber together with R.A. Millikan at the California Institute of Technology when he discovered the positron, a particle with the same mass as the electron but with the opposite charge.. .. Just a few years earlier, P.A.M. Dirac had presented his relativistic wave equation for electrons, which predicted the existence of particles with a charge opposite that of the electron. Originally, Dirac identified these as protons, but J. Robert Oppenheimer and others showed that the predicted particles must have the same mass as the electron and hence must be distinct from the proton. Anderson had discovered precisely the particle required by the Dirac theory, the antiparticle of the electron, the positron (Cahn and Goldhaber 2009, p. 5), footnote omitted).[2]

True enough. And good enough for textbook purposes. But lifeless in its avoidance of the revealing detail that makes science a distinctively human activity subject to mistake, contentious disagreement and serendipity. How then could Anderson's "golden event" have emerged—and was it really all that golden?

3.2 Anderson's Positive Electron

Anderson's discovery of the positron was in many respects the *accidental* byproduct of an experimental venture conceived and designed with a very different aim in mind. After having just completed his Ph.D. at the California Institute of Technology, Anderson was recruited by Robert Millikan to conduct experimentation using a cloud chamber that Anderson had designed and constructed in order to test and further develop certain theories that Millikan had proposed regarding cosmic rays. As recounted by Anderson:

> The aim of the experiment that led to the discovery of the positron was simply to measure directly the energy spectrum of the secondary electrons produced in the atmosphere and other materials by the incoming cosmic radiation which at that time (1930) was thought to consist primarily of a beam of photons or gamma rays of energies of the order of several hundred millions of electron volts. Although there was no experimental evidence as to the detailed interactions between such a beam of high-energy photons and matter, it was presumed from

[2]Other textbook statements are: "What may have been a fatal defect in 1930, however, turned into a spectacular triumph in late 1931, with Anderson's discovery of the *positron*, a positively charged twin for the electron, with precisely the attributes Dirac required" (Griffiths 2008, p. 21). Griffiths also uses photo from Anderson (1933b) and "The discovery of the positron by Anderson in 1933, with all the predicted properties, was a spectacular verification of the Dirac prediction" (Martin 2009, p. 8).

> experiments at lower energies that the dominant mechanism would be the production of high-energy secondary electrons by the Compton process. ... . (Anderson 1961, p. 825).[3][4]

Anderson's cloud chamber was unique at the time because of the very strong magnetic field that could be produced of up to 15,000 gauss.[5] The effect of the magnetic field is such that a charged particle would be "deviated into the arc of a circle" where the rotation was "clockwise for a particle of negative charge, and counter-clockwise for one of positive charge (Anderson 1932a, p. 406). But such deviations by themselves do not determine the charge of a particle because an observed arc could be interpreted as resulting from either a clockwise or counterclockwise motion. Therefore:

> The sign of the charge can be ascertained *only if* the direction of motion of the particle is known (Anderson 1932a, p. 406, emphasis added).

We highlight this experimental limitation because of its importance for the ultimate discovery of the positron. The limitation, however, was at least temporarily mitigated by the following working assumption made by Millikan and Anderson.

> It is assumed here that the particles are traveling downward through the chamber. The small degree of scattering to be expected for high-energy particles, combined with the known fact that the rays come in from above, appears to justify this assumption (Anderson 1932a, p. 406).

In other words, that while the assumption of downward motion is not entirely true because of scattering, the rarity of such events should have only minor effect. Elaborating further on such scattering, Anderson noted that:

> A possibility to be borne in mind is that, in rare cases, the tracks of curvature that indicate positives might be in reality electrons scattered backwards by the material underneath the chamber and are traversing it from bottom to top. Precise data on the specific ionization of the low-energy positives will distinguish, however, between downward positives and upward negatives (Anderson 1932a, p. 418).

The restriction here to "low-energy positives" is due to the fact that:

> *For a given velocity*, protons and electrons ionize the same, and for high energies where the velocities of both the electrons and the protons are of the order of the velocity of light it becomes impossible to distinguish electrons from protons by their ionization (p. 418, emphasis added).

There was, however, a window of opportunity in cases where lower energies were involved, because "at lower energies where the proton velocities are appreciably less than the velocity of light. .. protons [will] show an appreciably greater specific ionization than electrons *of the same energy* (p. 318, emphasis added)." Thus, in a

---

[3]For more on how and why Millikan coopted Anderson to engage in such cosmic ray research see De Maria and Russo (1985, pp. 238–239).

[4]The Compton process is the scattering of high-energy γ rays or X-rays from atomic electrons. Such a process will produce a high-energy electron.

[5]For a detailed description of the apparatus see Anderson (1933b, 407–408).

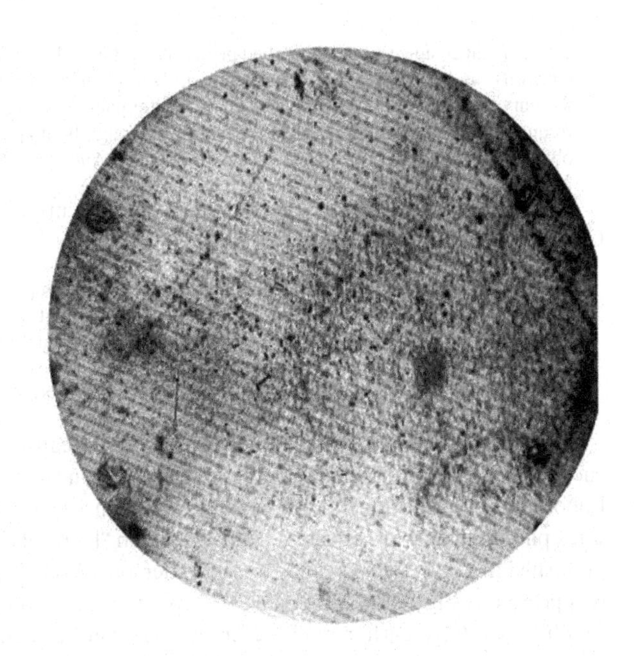

Fig. 3.2 Magnetic field, 17,000 gauss. A pair of associated tracks; at the left an electron of 120×10^6 V energy, at the right a proton of 130×10^6 V energy [where] the 130×10^6 V-proton is expected to ionize about 2½ times as heavily as the electron. *Source* Anderson (1932a)

case where a proton and energy are of the same or similar energy, the proton will have a smaller velocity because of its greater mass. And because of its smaller velocity, the proton will create greater ionization along its path.[6] For an example of what this looks like in practice see (Anderson 1932a, p. 408, Fig. 5) here reproduced as Fig. 3.2 where the proton track is distinguished by its (downward) counterclockwise track of greater ionization density.

In a nutshell, the analysis of the experimental data proceeded on the assumption that the particles (both negative and positive) were travelling downward and while distinguishing negative from positive was not possible in cases where high energies were involved they could be distinguished—or so it was thought—when low energies and correspondingly lower velocities were involved. But as Anderson would later note such low energy "was only true for a small fraction of the events (Anderson 1961, p. 825)." Still, while small, not zero.

We would be remiss, however, if we didn't bring to the reader's attention that there was an important implicit assumption made at this stage of the inquiry by Millikan and Anderson, namely, that there were only two kinds of charged particles: electrons and protons. The question is, what was the experimental evidence that convinced them otherwise?

The initial experimental results were successful with regard to their original purpose, and were, in fact, "dramatic and unexpected" (Anderson 1961, p. 825) insofar as they disproved the Compton process as an explanation of cosmic ray interaction. As summarized by Millikan and Anderson:

[6]For a more extensive description of these relationships see Blackett and Occhialini (1933, 703–704); and for the underlying theoretical basis see (Blackett 1932).

> The incident cosmic rays are absorbed primarily by the nucleus, rather than by extra-nuclear electrons, as heretofore generally assumed. This is shown by the new observation that the curvature of the tracks produced by a 17,000 gauss magnetic field corresponds more frequently to positive than to negative particles, though both appear. Positive particles can obviously come only from the nucleus (Millikan and Anderson 1932, p. 325).

And as later summarized by Anderson with a more explicit reference to the Compton process:

> The first result of the experiment was to show that the Compton process did not play an important role in the absorption of cosmic radiation, but that instead some new processes, presumably of a nuclear type, were operative. This was brought out by the fact that about half of the high-energy cosmic-ray particles observed were positively charged and therefore could not represent Compton electrons. (Anderson 1961, p. 825).

In November of 1931 Millikan gave pre-publication lectures describing this success and its experimental basis at the Institut Poincaré in Paris and the Cavendish Laboratory in Cambridge (Millikan and Anderson 1932, p. 325). Eleven cosmic-ray track photographs were presented, among which "[o]ne of the best" had already been published in *The Science News-Letter* (Anderson 1931). This photograph eventually reappeared as Fig. 5 in (Anderson 1932, p. 408) and is reproduced here as Fig. 3.2 to which we have earlier made reference. As can be seen there are two tracks of opposite curvature where the left track is identified (on the assumption that both were moving downward) as an electron and the right as a proton where the energies were determined to be respectively 120 and 130 million volts. And since the energies are so nearly equal, the ionization path of the proton should be around three times as heavy as that of the electron—which is what it clearly was. "Hence, we had no doubt whatsoever it was a proton (Millikan 1935, p. 328)." But as became evident later, this photograph was something of a black swan, since the ionization tracks of the apparent protons in other photographs were nowhere so much greater—if at all. In this regard, Skobeltzyn perceptively noted at the time that:

> The positive tracks on pictures demonstrated by Millikan did not differ essentially from electron tracks on the same pictures, with energies of about 50 MeV. Millikan and his audience overlooked this inconsistency (Skobeltzyn 1983, pp. 114–115).

On reflection, and after Millikan's otherwise triumphant return home, Millikan and Anderson also had cause for doubt especially with regard to what later appeared in Anderson's 1932 paper as Fig. 11 (Anderson 1932a, p. 411) and here reproduced as Fig. 3.3.

> The photograph shown in [Fig. 3.3]. .. bothered us very much, and Dr. Anderson and I discussed it at great length. The positive particle, to the right, has only a little curvature, and from this curvature we concluded that it was a proton of 450 million electron-volts of energy, while the track to the left was computed to be an electron of 27 million electron-volts. Now we knew that protons of more than a billion (10^9) electron-volts of energy would give an ionization along its track scarcely distinguishable from the ionization produced by a free electron, but at 450 million electron-volts the ionization due to a proton should be from 1½ to 2 times that due to an electron of 27 million electron-volts, and *yet no trace of a difference in ionization between the right- and the left-hand tracks of [Fig. 3.3] was discernable* (Millikan 1935, pp. 328–329).

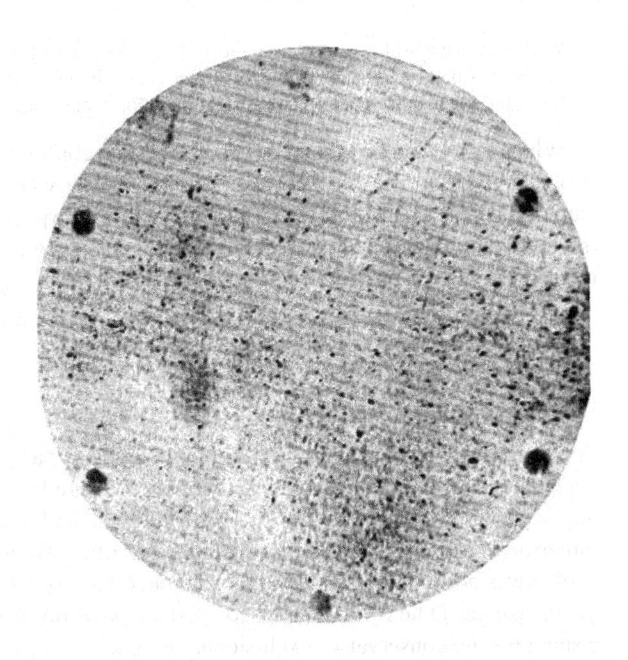

Fig. 3.3 Magnetic field, 12,000 gauss. An electron of 27×10^6 V energy and probably a proton of 450×10^6 V energy though the curvature here is small. A slight scattering is apparent. *Source* Anderson (1932a)

In response in part to this discrepancy, Anderson began to execute on his earlier promise to show that "[p]recise data on the specific ionization of the low-energy positives will distinguish. .. between downward positives and upward negatives (Anderson 1932a, p. 418)." And in this connection, we remind the reader that Anderson had earlier warned:

> A possibility to be borne in mind is that, in rare cases, the tracks of curvature that indicate positives might be in reality electrons scattered backwards by the material underneath the chamber and are traversing it from bottom to top (Anderson 1932a, p. 418).

Indeed, this possibility came to the fore when "precise data on the specific ionization of the low-energy positives" and its subsequent analysis revealed a troubling anomaly, namely, that the detected ionization levels indicated that the mass values for the (relatively slowly moving) protons was only of an order of magnitude on par with that of electrons.

> The first thing that came immediately out of the pictures was a set of high-energy particles of unit electric charge – roughly half positive and half negative.. .. Some of these particles, the positive ones, were moving slowly enough so they should have, if they had been protons, exhibited an increase in ionization, which they did not do (Anderson 1979, p. 35).

The reader may wonder how this deficit in ionization was determined. As Anderson explained:

> [Y]ou could also count droplets under a microscope. But it was essentially a visual thing. You could just look at it and see that you had electrons (p. 35).

But there was an important caveat to the efficacy of such an "eyeballing" determination:

Well, they were still, in most cases, [of] high enough energies so the effect wasn't a big effect. You had to worry about instrumental uncertainties and intensity of the light in the film. But you had the electron tracks right there for comparison (pp. 35–36).

Whatever Anderson's worries had about such confounding uncertainties became increasingly less pressing because "[a]s more data were accumulated, however, a situation began to develop which had its awkward aspects in that *practically all* of the low-velocity cases were particles whose mass seemed to be too small to permit their interpretation as protons (Anderson 1961, p. 826, emphasis added)."

The alternative interpretations in these cases were that these particles were either electrons (of negative charge) moving upward or some unknown lightweight particles of positive charge moving downward (p. 826).

Anderson reports that "[i]n the spirit of scientific conservatism we tended at first toward the former interpretation, i.e., that these particles were upward-moving negative electrons (p. 826)." Conservative to be sure because it closed the door for any admission of a new type of fundamental particle, but given the conflict with the hitherto working hypothesis that cosmic ray particles tend to move downward, this choice was admittedly "not a very good one (Anderson 1979, p. 35)." Surprisingly given his later fame for having discovered the positron, Anderson at this stage made a stand for the conservative exclusionary choice:

And I had discussions with Dr. Millikan about this and I said, "You wouldn't expect it, but they *must* be electrons that are going up," because the tracks weren't heavy enough to be interpreted as protons (Anderson 1979, p. 35, emphasis added).

Millikan disagreed:

[Which] led to frequent, and at times somewhat heated, discussions between Professor Millikan and myself, in which he repeatedly pointed out that everyone knows that cosmic-ray particles travel downward, not upward, and therefore these particles *must* be downward-moving protons (Anderson 1961, p. 826, emphasis added).

In order to resolve the dispute about what "*must*" be the case, "[a] lead plate was inserted across the center of the chamber in order to ascertain the direction in which these low-velocity particles were traveling" (Anderson 1961, p. 826).[7] Anderson was apparently exercising a bit of poetic license here because an inserted lead plate had already been used for some of the earlier cosmic ray photographs reported in Anderson (1932a). And while most of the lead plate photographs dealt with absorption by the plate, in two of the photographs a cosmic ray particle "of uncertain sign" (because of the lack of discernable ionization curvature) enters and exists the plate with however a discernable angular "deflection (see Figs. 22 and 23, Anderson 1932a, p. 417). About these photographs, Anderson had this to say:

Experiments are now in progress to study the scattering of cosmic particles in lead. Figures 22 and 23 show particles traversing 6.0 mm of lead, the angle of scattering being in each case readily measurable (p. 420).

---

[7]As similarly stated in Anderson (1965, p. 366, 1966, pp. 20–21, 1979, p. 36) and Anderson and Anderson (1983, p. 140).

Note, however, that here the interest is in "the scattering of cosmic particles" and not in the determination of the directionality of their motion. That would require lower velocity and as a result detectable curvature of the ionization path.

Happily for Anderson, the use of a lead plate in order to resolve the question of directionality resulted in the photographic capture of what was clearly an upwardly moving particle of positive charge. This was the "golden" and many times reproduced Fig. 1 of (Anderson 1933a, 492).[8] A later reminiscence about this is worth repeating because of its relevance for questions of replication.

> LYLE: You had this one experiment where it was clear that this was a positive electron. *Did you attempt to repeat this*? How difficult was it to repeat that?
>
> ANDERSON: Well, that was by far the best photograph. But there were other things—cosmic ray particles, as was often observed, came in showers, groups of particles. Some of these were usually of lower energy than when they came singly like in most cases. And there it was more striking; if the positive particles were protons, they had low enough energy so they should have clearly given a heavier track than they did. So, there was that kind of evidence before this one that I could call the *clinching picture*. We used to talk about positive electrons before that picture, but never really took it seriously (Anderson 1979, p. 36, emphasis added).

So it was the "clinching picture" that decisively turned the tide in favor of Millikan's resolution of the up-down dilemma. Moreover, as suggested by Anderson's response to Lyle's question, whatever concerns that may have existed about replication were alleviated by the earlier shower photographs.

There was, however, a problem. The postulated positron entered the plate from below and not from above as would be expected given that cosmic-ray particles go down. So ironically, "despite the strong admonitions of Dr. Millikan that upward-moving cosmic-ray particles were rare, this indeed was an example of one of those very rare upward-moving cosmic-ray particles" (Anderson 1961, p. 826). Thus, was the improbable recruited to save the wayward positive charge.

So much for the tortuous path to discovery. How though was it to be packaged for publication? With the not unimportant issue of priority in mind Millikan pressed for a brief and quick publication in *Science* and not the *Physical Review*. Anderson agreed and his report was indeed brief, contained no photographs, attracted little attention and came to be regretted.[9] But it did establish priority—but only, we think,

---

[8] Anderson, however, did initially have some doubt about its "golden" status and as recounted by Luis Alvarez who was at Cal Tech at the time, "the first thing [Anderson] thought was, "Those Cal Tech jokesters have reconnected wires to my magnet and have reversed its polarity." And the reason that he thought that was that it hadn't been so much earlier when the Caltech students had reconnected the wires on the wind tunnel motors that drove the big fans and they went backwards.... But then [Anderson] looked [at] pictures just prior to the positron picture and just after that and the electrons had the right sign, and it was clear that no one could have disconnected the thing when it was hot, and reconnect it; so the field must have been in the same sense all the time (Alvarez 2014, p. 16).".

[9] For Anderson's account of the decision to rush publication in *Science* (Anderson 1932b) see Anderson (1966, pp. 23–24, 1979, p. 37). Oddly enough Anderson did manage shortly thereafter to get a short letter published in the *Physical Review* which however made no mention of a positive-electron discovery but summarized evidence obtained regarding "the frequent occurrence of associated tracks" that could not be explained by "a simple binary collision." The tracks,

because it was backed up by the real deal in the form of Anderson's 1933 report in the *Physical Review* (Anderson 1933a).[10] And it is to that publication that we now turn.

After a few brief introductory remarks, and posting Fig. 3.1 front and center, Anderson went directly to the experimental proof that the particle could not have been a proton. The proof is surprisingly straightforward. First, if the upward moving particle had the mass of a proton then its energy is determined by the curvature and has a value of 300,000 volts. But such a particle would have a tracking *range* far less than that observed. As succinctly stated by Anderson:

> [A] proton of that energy according to well established and universally accepted determinations has a total range of about 5 mm in air while that portion of the range actually visible in this case exceeds 5 cm without a noticeable change in curvature (p. 491).

Therefore, the particle observed cannot be a proton. QED. Plain, simple and "inevitable" (p. 491).

Anderson's reliance on the relationship between the particle's energy and the range of its ionization track is noteworthy because range had not hitherto been a consideration. But here it was employed with decisive effect. Well not quite since there was the possibility that the entry and exit tracks were not what they appeared to be and that "[at] exactly the same instant (and the sharpness of the tracks determines that instant to within about a fiftieth of a second) two independent electrons happened to produce two tracks so placed as to give the impression of a single particle shooting through the lead plate" (p. 491) But this possibility "was dismissed on a probability basis" because a track of the order of curvature observed occurred "only once in some 500 exposures," and there was thus "practically no chance at all that two such tracks should line up in this way" (p. 491).

Dismissed as "completely untenable" was any assumption that the tracks were produced by an electron coming from above. Untenable because such an interaction with the lead plate would have caused an increase in energy from 20 million (at entrance) to 60 million volts (at exit).

This left the more creative possibility that "[a] photon, entering the lead from above, knocked out of the nucleus of a lead atom two particles, one of which shot upward and the other downward. But in this case the upward moving one would be a positive of small mass" in which case there would still be the occurrence of a particle that was not a proton (p. 491).

Assuming all this and that "once was enough" to show that the particle observed was not a proton, there still remained the question of if not a proton, then what? What was the size of its positive charge, and what was its mass? Here the argumentation is less clear cut because "[i]t is possible with the present experimental data only to

---

however, were identified on the basis of their "point of origin" or "on the basis of curvature, range and specific ionization" as electrons or in two cases as "either electrons or protons (Anderson 1965, pp. 368–369).".

[10]As had been the case with the initial report in *Science*, this paper while more extensive and complete with photographs was evidently also somewhat rushed as evidenced by its lack of an author prepared abstract. Thus the published abstract was prepared by and attributed to "Editor.".

assign rather wide limits to the magnitude of the charge and mass of the particle" (pp. 491–492). With respect to charge there were two such determinations of the "wide limits." First:

> The specific ionization *was not in these cases measured*, but it appears very probable, from a knowledge of the experimental conditions and *by comparison* with many other photographs of high- and low-speed electrons taken under the same conditions, that the charge cannot differ in magnitude from that of an electron by an amount as great as a factor of two (p. 492, emphasis added).

But Anderson had in his arsenal more than just this sort of informal "eyeballing." Namely, a low level theoretical but empirically based argument based on the loss of energy due to the penetration and passage through the plate that led to "an upper limit to the charge less than twice that of the negative electron." For the details see (pp. 492–493). But if indeed the charge of the positron is less than twice that of an electron, then assuming that charge does not come in fractional amounts, it follows that the charge of the positron is "very probably" that of the electron:

> It is concluded, therefore, that the magnitude of the charge of the positive electron which we shall henceforth contract to *positron* is very probably equal to that of a free negative electron which from symmetry considerations would naturally then be called a *negatron* (p. 493, emphasis added).[11]

With respect to the mass of the positron, the best that Anderson could muster was an upper limit of twenty times the mass of the electron.

> The magnitude of the proper mass cannot as yet be given further than to fix an upper limit to it about twenty times that of the electron mass. If Fig. 1 represents a particle of unit charge passing through the lead plate then the curvatures, on the basis of the information at hand on ionization, give too low a value for the energy-loss unless the mass is taken less than twenty times that of the negative electron mass (p. 493).

Three additional photographs were included in Anderson's 1933 discovery paper, though indicative of their supporting status Anderson's commentary is restricted to the captions. The first (Fig. 2) shows an electron and a positron apparently originating from a common source, proceeding down from the bottom of the plate. The second (Fig. 3) (here reproduced as Fig. 3.4) shows six particles projected from the wall of the chamber including two pairs apparently from a common source. The third (Fig. 4) shows a positron (though with a very faint ionization trial) coming from above and existing from below. While not stated explicitly this presumably was included to dispel any doubts about the positron coming from below in Fig. 1. For

[11]This marks the first published appearance Anderson's use "positron" as the name for the newly discussed positive electron. The abstract which was prepared by the editor of *Physical Review* also states that "[t]hese particles will be called positrons." This has suggested to some that "positron" originated with the *Physical Review* editor. But this is mistaken since the name was first proposed by Watson Davis, an editor at *Science Service* (a Washington D.C. based science publisher), who had earlier sent Anderson a telegram suggesting that "positron" and "negatron" be used to respectively denote the positive electron and the (negative) electron. After contemplating Davis' suggestion over a game of bridge Anderson wired back his agreement. See Anderson (1966, pp. 26–27). And while "positron" caught on "negatron" did not.

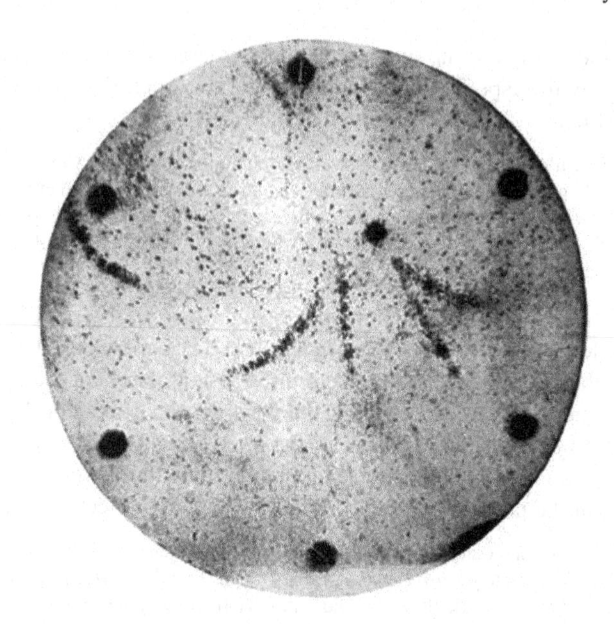

Fig. 3.4 A group of six particles projected from a region in the wall of the chamber. The track at the left of the central group of four tracks is a negatron of about 18 million volts energy (Hp = 6.2104 gauss-cm) and that at the right a positron of about 20 million volts energy (Hp = 7.0 × (10^4 gauss-cm). Identification of the two tracks in the center is not possible. A negatron of about 15 million volts is shown at the left. This group represents early tracks which were broadened by the diffusion of the ions. The uniformity of this broadening for all the tracks shows that the particles entered the chamber at the same time. *Source* Anderson (1933a)

future developments, Fig. 3 was the most interesting because it included two showers of the sort which were held to be most indicative of how positrons were produced by cosmic and gamma ray collisions.

With the existence of the positron and a rough measure of its mass and charge established, Anderson closes by considering how it is that positrons are produced. The most relevant consideration is the close association between positrons when they appear with companion electrons.[12] Thus:

> From the fact that positrons occur in groups associated with other tracks it is concluded that they must be secondary particles ejected from an atomic nucleus. (p. 494)

As a first step in the development of such a theory of production Anderson undertook a more precise determination of this association. In short, and in its most basic form, the idea was to determine how often positrons appeared along with their sibling electrons. This involved (1) the selection of representative episodes where the corresponding ionization tracks were well defined, and (2) the development of a more accurate method (involving microscopic examination) of determining ionization levels.

[12]One irony to be noted here is that the "golden" encounter captured in Fig. 3.1 consisted of only a single positron which had evidently separated from its electron sibling and then bounced back into and through the lead plate.

Anderson must have already been working along these lines since these developments were reported only a few months later in Anderson (1933b). There were two important results. First, that the "differences in specific ionization of 20% could be readily detected" (p. 408). And second that subject to variation in the particle energies there was a "general [statistical] symmetry in occurrence between the positives and negatives is in accord with the view that in the primary absorption process of the incident cosmic-ray beam positives and negatives are formed in pairs perhaps as governed by Dirac's theory of electrons" (p. 411).[13] And here we have Anderson's first reference to Dirac. But very brief and with no further elaboration.

In a later publication, Anderson drew out the consequences of these results for a more accurate determination of the charge and mass of the positron than had been earlier achieved.

> More recent measurements of the specific ionisation of the positives and negatives for both high and low speed particles, *by actual ion-counts on the tracks* in the magnetic field, showed the specific ionisation of the positives and the negatives to be equal to within 20 per cent. This fixes the limits of difference between the positives and negatives with regard to charges and masses at 10 per cent and 20 per cent respectively (Anderson 1934. p. 313, emphasis added).

This was a considerable improvement over Anderson's earlier determination, achieved by an informal "eyeballing" analysis of "an upper limit to the charge less than twice that of the negative electron" and a mass value of "less than twenty times that of the negative electron mass (Anderson 1933a, p. 493)."

Finally we note that "once was enough" not only for Anderson to embark on a more serious determination of the association of positron and electron tracks and a more accurate determination of the mass and charge of the positron, but it was also enough for Anderson, along with many others, to get into the business of producing positrons by gamma rays. For Anderson's own appraisal on this matter see Anderson (1934). But rather than proceeding along these lines, we'll move on to a consideration of the more or less simultaneous discovery of the positron by Paul Blackett and Giuseppe Occhialini.

3.3 Blackett and Occhialini: Less Precise, but More Showers

Blackett and Occhialini became aware of Anderson's announcement of the discovery of the positron made in *Science* and soon thereafter submitted their own account which unlike Anderson's rather short report contained extensive analysis as well as fifteen photographs (Blackett and Occhialini 1933). Thus, while Blackett and Occhialini may not have been the first to report they were certainly the first to have published cloud-chamber photographs with identified positron ionization tracks. But

[13]There was, however, a more extensive but still relatively brief discussion of Dirac in the later (Anderson 1934, p. 314).

as we shall see, none were nearly so "golden" as Anderson's soon to be published Fig. 3.1.

Before proceeding with a more formal examination, we would like to draw attention to what, we expect, many readers will find striking when comparing the photographs of Blackett and Occhialini with those provided by Anderson. Namely, that it's as if Blackett and Occhialini, and Anderson occupied different worlds. Blackett and Occhialini's photographs were typically described as being "beautiful." And so they were with their gloriously profuse and explosive "showers" of cosmic ray particles. But there was no such praise for Anderson's photographs. See, for example, the Blackett and Occhialini photographs together reproduced here as Fig. 3.5. The best that Anderson had to offer in the shower department was the rather dowdy photograph reproduced earlier as Fig. 3.4.

Why the difference? Answer: 17,000 gauss—the difference between the magnetic field imposed by Anderson (of up to 20,000) and the considerably lesser field strength

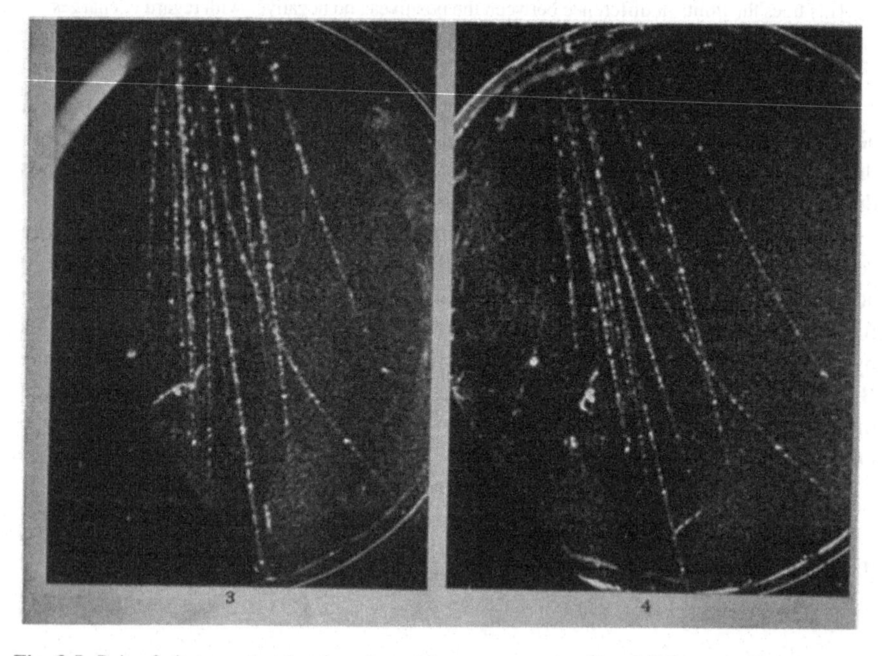

Fig. 3.5 Pair of photographs showing about 16 separate tracks. H = 3100 gauss. The divergent point of the shower is again in the copper coils. On the left are two negative electron tracks with $H\rho \sim 0.5 \times 10^5$ and so $E_e \sim 15$ MV. On the right are two tracks curved markedly to the right, which must be due to positive electrons, with $H\rho \sim 0.4$ and 1.5×10^5, and so $E_e \sim 12$ and 45 MV. Some of the other tracks are slightly curved, some one way, and some the other. Most of the nearly straight tracks seem to diverge from the same point, but the more bent ones probably diverge from a secondary radiant point lower down. *Source* Blackett and Occhialini (1933)

(of only 3000 gauss) mustered by Blackett and Occhialini.[14] Hence, the shower paths of Blackett and Occhialini's photographs exhibited smaller curvature and consequently were more narrowly confined. By contrast, Anderson's showers dispersed more quickly—with the resulting loss of aesthetic appeal. But that's just part of the answer. Equally important is the fact that because of the automatic triggering mechanism employed, Blackett and Occhialini had many more photographs at their disposal than Anderson who had to make do with random photographic bursts with fingers crossed that something of interest would be captured. Hence Blackett and Occhialini were more likely to capture the relatively rare shower events. For an analysis of the capture rates for showers see (Blackett and Occhialini 1933, pp. 709–711).

Of course, what matters is not aesthetic appeal, but the quality of the data, and here Blackett and Occhialini had to admit that:

> It is unfortunate that in our experiments the magnetic field was not sufficient to deflect appreciably the majority of the tracks, either those appearing singly or those in showers. But, as has already been mentioned, the measurements of Anderson. .. were made with much larger fields, and. .. show conclusively that about equal numbers are deflected either way, and this is also true of those tracks in our photographs which are appreciably deflected. (Blackett and Occhialini 1933, p. 707)

The consequences for the determinations of the mass and charge of the newly discovered positron will be examined below, but for now we note that Blackett and Occhialini preface their analysis with the disclaimer that:

> In this paper a preliminary and *mainly qualitative account* will be given of some of the more striking phenomena observed in the photographs, leaving almost all the details and the measurements for subsequent reports (p. 702, emphasis added).

Still, the deflections observed by Blackett and Occhialini were sufficient to support the conclusion that some of the particles observed of positive charge were not protons. And here unlike Anderson who began with his "clinching" photograph of the positron entering and exiting the lead plate, Blackett and Occhialini begin with their two best shower photographs here reproduced above as Fig. 3.5 (Blackett and Occhialini 1933, p. 723, no. 3 and 4). Their analysis begins with the reasonable assumption that:

> If a group of tracks diverge from some point or some small region of space, then there is a high probability, but no certainty, that any one particle did actually move away from this region (Blackett and Occhialini 1933, 705).[15]

Assuming that this is so allows for a determination of the charge polarity for tracks with appreciable curvature.

---

[14]For a description of the apparatus used by Blackett and Occhialini and its automatic triggering mechanism see (Blackett and Occhialini 1933, pp. 699–701).

[15]This assumption was more colorfully rendered by Anderson as applying to cases "where one had a group of, say, three particles clearly coming from an origin and therefore where the direction of motion was known, because one couldn't have particles that agreed to have an appointment at some place in the future" (Anderson 1966 p. 20).

> Now on photographs 3 and 4 the majority of the tracks are undeflected. .. ($H\rho > 10^6$), but there are some which are definitely curved and of these some are curved in one direction and some in the other. The appearance of the whole group strongly suggests that the particles have diverged downwards, and if this is assumed true, those bent to the left are negatively charged and those bent to the right are positively charged (Blackett and Occhialini 1933, pp. 707–706, footnote omitted).[16]

Blackett and Occhialini then state that "[o]n the right are two tracks curved markedly to the right, which must be due to positive electrons, with $H\rho \sim 0.4$ and 1.5×10^5, and so $E_e \sim 12$ and 45 MV" (p. 722). What they mean here is that in each of the photographs there are two such tracks where in each of the photographs the track of higher energy is to the left of the (less curved) track of lower energy. Blackett and Occhialini then calculate the anticipated range of these tracks on the assumption that they were due to protons and find that the observed range far exceeds those values.

> If these two tracks were due to protons their ranges could only be 0.2 cm. and 3 cm., in air, whereas their actual lengths in the chamber are about 12 cm. of air at N.T.P. So these tracks are certainly not due to protons, but to particles of much smaller mass.. .. (pp. 705–706, footnotes omitted).

As an important and here significant methodological point, we note that Blackett and Occhialini, as had Anderson, took advantage of the following feature of range as a determinate of mass.

> In rare cases a particle stops in the gas of the chamber so that its range can be measured, and even when this is not so, the knowledge that the range is greater than a definite amount is sometimes of *decisive* importance (p. 703, emphasis added).

We note also that Blackett and Occhialini were able to avoid the problem of distinguishing a downward positive from an upward electron by making use of their shower photographs which indicated a common source point from above.[17] Anderson, however, did not (at least initially) have such photographs with a clearly indicated origin and thus had to resort to his lead plate to decisively decide the matter.

There is a subtle difference between the way that Blackett and Occhialini and Anderson make use of ionization density. For Anderson, the procedure was to assume that because of the extended range of the track (in the lead plate photograph) decisively showed that the ionization path was not due to a proton. A comparison of the ionization densities due to electrons as compared with that of the newly discovered positron was then used to determine the mass and charge of the positron. Blackett and Occhialini, by contrast, used their mass determination to supplement the conclusion of their range argument. Thus they argued that "[t]he study of the ionization density along the tracks entirely confirms" that "these tracks are certainly not due to protons,

---

[16] $H\rho$ is the product of the magnetic field and the radius of the curvature of the ionization track.

[17] With regard to their shower photographs Blackett and Occhialini noted: "It is, of course, conceivable that some of these tracks are caused by negative electrons moving upwards, which only by chance pass through the region from which the other tracks appear to diverge. It is difficult to estimate this chance numerically, but the presence of these positively curved tracks is so common a feature of these showers that this explanation can hardly be maintained for them all (p. 706)." Note that this a probabilistic argument.

but to particles of *much smaller mass*" (p. 706, emphasis added). But because of the considerably smaller deflections involved Blackett and Occhialini had to make do with a "much smaller mass" as opposed to the more precise determinations obtained by Anderson. As further elaborated by Blackett and Occhialini:

> It is strikingly evident from their appearance that the ionization density along the positively curved tracks in the photographs mentioned is very nearly the same as that along the undeflected and negatively curved tracks. It is, unfortunately, *not possible to make a precise count* of either the primary or the total ionization along these tracks without special experiments, but *rough counts* have shown that the ionization density common to all these tracks is *approximately* that to be expected for fast electrons (p. 706, footnote omitted).

Accordingly:

> The only possible conclusion of the argument *both* from the range and from the ionization is that these tracks are due to positively charged particles with a mass *comparable* with that of an electron rather than with that of a proton (p. 706, footnote omitted, emphasis added).

Blackett and Occhialini were evidently aware that while "much smaller" and "comparable" were sufficient to supplement their range based argument, these "qualitative" appraisals had to be improved upon.[18] Thus, they went on to later emphasize that "[i]t will be a task of immediate importance to determine experimentally the mass of the positive electrons by accurate measurements of their ionization and $H\rho$ (p. 715)."

Blackett and Occhialini also had photographs of inserted plates, but these were problematic and by no means as convincing as Anderson's "clinching" photograph. In fact, according to their own appraisal only one, here reproduced as Fig. 3.6, was a serious contender. Given its subsidiary status, Blackett and Occhialini are brief in their analysis:

> Additional evidence, independent of the showers, for the existence of a positively charged particle with a mass comparable with that of an electron is found in photograph No. 7. In this, the particle passes through a 4 mm. lead plate. The track is seen to be more curved below the plate than above, so the particle must have moved downwards unless it be assumed that it gained energy in traversing the plate; hence it had a positive charge. Corresponding to the final $H\rho$ of 1.2×10^5 gauss-cm. it would only have had a range of 1.5 cm. in air if it had been a proton and it would have ionized more than 100 times as much as a fast electron. Its range is certainly greater than 5 cm. and it ionizes just like a fast electron (p. 707).

So as with Anderson, the argument turns on the fact that the calculated range of a proton ionization track is greatly exceeded by the range of the observed track. If this were the end of the matter Anderson's plate observation and that of Blackett and Occhialini would appear to be on par. Closer examination, however, indicates otherwise. First, as Blackett and Occhialini point out in their caption for the photograph, the "energy loss [is] too great [for normal absorption], but this is not unexpected since energy may well have been lost in the collision causing the deflection." While not

[18]Because Anderson submitted his 1933 report later (received Feb 28, 1933 and published March 15, 1993) than did Blackett and Occhialini (received Feb 7, 1933 and published March 3, 1933), Blackett and Occhialini were likely unaware of Anderson's more precise mass and charge determinations.

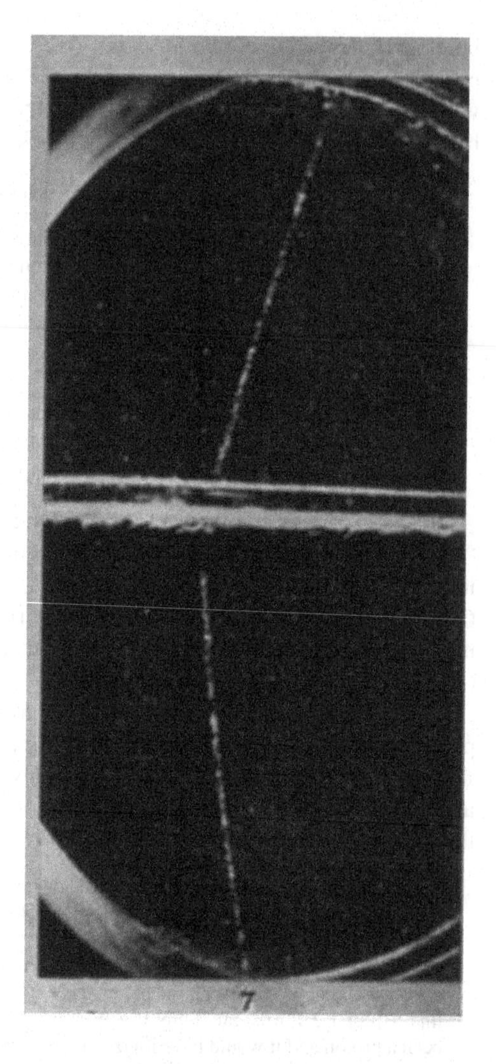

Fig. 3.6 Magnetic field = 2200 gauss. 4 mm. lead plate. A track showing deflection at plate and a greater positive curvature below than above. The direction must therefore be downwards, and hence the charge positive. Above plate $E_e \sim 60$ MV. Below, $E_e \sim 22$ MV. Energy loss too great again, but this is not unexpected since energy may well have been lost in the collision causing the deflection. *Source* Blackett and Occhialini (1933)

entirely clear, we think that what Blackett and Occhialini had in mind is the fact that the incoming and existing tracks are very nearly straight and at most of very small (though differing) curvature. Significantly the tangents at the entry points are not parallel (as they were in Anderson photograph) but rather display a pronounced and sudden angular change of direction presumably caused by some sort of secondary "collision" and consequent scatter in the plate.[19,20] Given this complication, it must

[19]This indicates that the particle lost energy. and was moving downward.

[20]Blackett and Occhialini did, in fact, attribute the discovery of the positron to Anderson. Interestingly, they cited Anderson's 1932 paper. "But it will be shown that it is necessary to come to the same remarkable conclusion that has already been drawn by Anderson* from similar photographs. This is that some of the tracks must be due to particles with a positive charge but whose mass is much less than that of a proton (Blackett and Occhialini 1933, p. 703)

be said that Blackett and Occhialini's plate photograph is not nearly as compelling as Anderson's. Which is why Blackett and Occhialini gave it only supporting status for their more convincing shower photographs.

After making their case for the existence of the positron, Blackett and Occhialini, as had Anderson,[21] went on to discuss the relevance of this discovery for the question of the "mechanism" of the production of cosmic ray particles (see Blackett and Occhialini 1933b, pp. 711–714). Surprisingly, Blackett and Occhialini make no mention of Dirac's theory in all of this. And in fact, Dirac and his theory are not mentioned in Blackett and Occhialini's (1933) report until nearly the very end. What explains this?

3.4 Dirac's Theory Makes Its First Appearance

Before proceeding, we should note that Anderson avoided any mention of Dirac in his 1933 paper. He was quite emphatic when he later insisted that Dirac's theory played no role in his experimental endeavors.

> [W]ith respect to the positron, it has often been stated in the literature that the discovery of the positron was a consequence of its theoretical prediction by Dirac, but this is not true. The discovery of the positron was wholly *accidental*. Despite the fact that Dirac's relativistic theory of the electron was an adequate theory of the positron, and despite the fact that the existence of this theory was well known to nearly all physicists, it played no part whatsoever in the discovery of the positron.
>
> The aim of the experiment that led to the discovery of the positron was simply to measure directly the energy spectrum of the secondary electrons produced in the atmosphere and other materials by the incoming cosmic radiation. .. . (Anderson 1961, p. 825, emphasis added).[22]

By contrast Dirac at least got some billing from Blackett and Occhialini if only in a minor supporting role in the last act. But before discussing Dirac's late appearance, we'll give a brief rendition of the theoretical underpinning involved in Dirac's prediction of the positron, and what it was that the discovery of the positron could be taken to have confirmed.

---

[21] Blackett and Occhialini did, in fact, attribute the discovery of the positron to Anderson. Interestingly, they cited Anderson's 1932 paper. "But it will be shown that it is necessary to come to the same remarkable conclusion that has already been drawn by Anderson* from similar photographs. This is that some of the tracks must be due to particles with a positive charge but whose mass is much less than that of a proton (Blackett and Occhialini 1933, p. 703).

[22] See also: "Yes, I knew about the Dirac theory. I didn't know it in its details, but I knew that there were in this relativistic equation these negative energy states, which he interpreted ... Or the absence of a particle in the negative energy state he interpreted as a proton, but with difficulties in the theory that apparently one couldn't have a different mass for a whole particle than the particle that was in the positive energy states. But I was not familiar in detail with Dirac's work. I was too busy operating this piece of equipment to have much time to read his papers (Anderson 1966, p. 21, ellipsis in original).

In 1928 Dirac published a landmark paper that contained the derivation of the equation that bears his name (see Dirac 1928a, 1928b). Briefly stated, the aim was to develop a consistently relativistic generalization of the Schrödinger and Klein-Gordon equations regarding the state functions of a quantum-mechanical system. While the positive energy solutions for the Dirac Equation led to many outstandingly successful predictions and explanations, there was a formal glitch in the form of negative energy solutions. Such solutions were mathematically unavoidable because solving the relativistic equation for energy involved the determination of a square root.[23]

As discussed by Dirac, negative solutions could be ignored in the context of classical physics because "one had only to assume that the world had started off with all particles in positive energy states [and that] they would then always stay in positive energy states." But "[w]ith the development of quantum theory. .. jumps can take place discontinuously from one energy level to another. If we start off with a particle in a positive energy state, it may jump into a negative energy state (Dirac 1983, p. 50)." Moreover, Dirac's Equation exacerbated the problem of what to do with the negative solutions because now there were four solutions, two of positive energy which accounted for the properties of electrons with opposite spin, and two of negative energy. Thus, "because the negative energy states cannot be avoided, one must accommodate them in the theory." But how? Dirac's answer was to develop "a new picture of the vacuum" (Dirac 1983, p. 51). This new picture made its debut public appearance in Dirac (1930). The initial move took account of the fact, as recognized by Weyl and others, that the motion of a wave packet of negative energy states, when formally conceived, has the following property:

> [It] is a possible trajectory for an ordinary electron (with positive energy) moving in the electromagnetic field with reversed sign, or for an electron of charge + e (and positive energy) moving in the original electromagnetic field. Thus *an electron with negative energy moves in an external field as though it carries a positive charge* (p. 361).

This suggests that "that a negative-energy electron *is* a proton (p. 362)." But as shown by Dirac, such an identification leads to several paradoxes including a violation of "the law of conservation of electric charge (p. 362)." So, the identification won't do as it stands. In order to resuscitate the proposed identity in an acceptable form, Dirac introduced the "hole" hypothesis. His introductory statement of the hypothesis is remarkably concise for such a dramatic and revolutionary proposal.

> The most stable states for an electron (i.e., the states of lowest energy) are those with negative energy and very high velocity. All the electrons in the world will tend to fall into these states with emission of radiation. The Pauli exclusion principle, however, will come into play and prevent more than one electron going into any one state. Let us assume there are so many electrons in the world that all the most stable states are occupied, or, more accurately, that *all the states of negative energy are occupied except perhaps a few of small velocity*. Any electrons with positive energy will now have very little chance of jumping into negative-energy states and will therefore behave like electrons are observed to behave in the laboratory.

---

[23]For more details on the Dirac Equation and the positive and negative energy solutions see Weisskopf (1983) and Pashby (2012, pp. 443–445).

> We shall have an infinite number of electrons in negative-energy states, and indeed an infinite number per unit volume all over the world, but if their distribution is exactly uniform we should expect them to be completely unobservable. *Only the small departures from exact uniformity, brought about by some of the negative-energy states being unoccupied, can we hope to observe* (p. 362).[24]

After considering some formally analogous situations involving "X-ray levels in an atom with many electrons," Dirac moves on to assert that:

> These [vacant states or "holes"] will be things of positive energy and will therefore be in this respect like ordinary particles. Further, the motion of one of these holes in an external electro-magnetic field will be the same as that of the negative-energy electron that would fill it, and will thus correspond to its possessing a charge + e. We are therefore led to the assumption that *the holes in the distribution of negative- energy electrons are the protons.* When an electron of positive energy drops into a hole and fills it up, we have an electron and proton disappearing together with emission of radiation (pp. 362–363).

After showing that the "paradoxes" caused by prematurely identifying the negative energy solutions with protons are avoided by means of his hole theory, Dirac draws attention to a major ontological benefit:

> We require to postulate only one fundamental kind of particle, instead of the two, electron and proton, that were previously necessary. The mere tendency of all the particles to go into their states of lowest energy results in all the *distinctive* things in nature having positive energy (pp. 363–364).

Legerdemain extraordinaire! While far-fetched for many, Weyl and Oppenheimer took the proposal seriously and convincingly demonstrated that Dirac's holes, whether occupied by negative energy electrons or interloping protons, would have to have the same mass.[25] And to further put the nail in the coffin, Oppenheimer went on to demonstrate that on Dirac's theory "the mean life time for ordinary matter [would be] of the order of 10^{-10} s" (Oppenheimer 1930, p. 563).

> Thus, we should hardly expect any states of negative energy to remain empty. If we return to the assumption of two independent elementary particles, of opposite charge and dissimilar mass, we can resolve all the difficulties raised in this note, and retain the hypothesis that the reason why no transitions to states of negative energy occur, either for electrons or protons, is that all such states are filled (p. 563).

In other words, Oppenheimer's proposal is to return to the ontology of electrons and protons, and where *all* negative energy states are filled by either electrons or protons. Dirac, however, was not to be so easily deterred. Though he accepted Oppenheimer's proposal that all states would be so filed, he left open the possibility that if an open occupancy should become available (because of some perturbation) then that unoccupied hole would manifest itself as a positive-electron, with however a very short lifetime. Thus, we have in 1931 the final step in the genesis of Dirac's prediction of the positron. In his own words:

---

[24] Note here the opportunistic application of the Pauli exclusion principle to the entire Dirac vacuum, which became known as the "Dirac Sea." This application of the principle later came under scrutiny. See Pashby (2012, pp. 455–456).

[25] See Weyl (1932, p. 263) and Oppenheimer (1930).

Following Oppenheimer, we can assume that in the world as we know it, *all*, and not merely nearly all, of the negative-energy states for electrons are occupied. A hole, if there were one, would be a new kind of particle, unknown to experimental physics, having the same mass and opposite charge to an electron. We may call such a particle an anti-electron. We should not expect to find any of them in nature, on account of their rapid rate of recombination with electrons, but if they could be produced experimentally in high vacuum they would be quite stable and amenable to observation. (Dirac 1931, p. 61).

On a first reading this is somewhat confusing because Dirac initially appears to accept, without qualification, Oppenheimer's proposal that "*all*, and not merely nearly all" negative-energy states are occupied. But he then opens the door to the possibility of unoccupied holes which while unlikely to be found "in nature" may, however, be "produced experimentally" where these unoccupied holes will manifest themselves as an "anti-electron." With regard to such a production of an anti-electron Dirac suggested that an experimentally facilitated collision of two gamma rays would be sufficient.

An encounter between two hard γ-rays (of energy at least half a million volts) could lead to the creation simultaneously of an electron and anti-electron, the probability of occurrence of this process being of the same order of magnitude as that of the collision of the two γ-rays on the assumption that they are spheres of the same size as classical electrons. This probability is negligible, however, with the intensities of γ-rays at present available (p. 61).

As history would reveal Dirac proposed one too many gamma rays. One was enough. But that was only discovered and understood after the discovery of the positron by means of a cosmic-ray interaction. When later asked about Dirac's influence on Blackett and Occhialini's cosmic ray experiments Blackett indicated (as reported by J.L. Heilbron) that:

Dirac worked very closely with them; in fact, he was often at the laboratory. When asked how long they had known about Dirac's theory, Blackett replied he wasn't quite certain, but that it didn't matter anyway *because nobody took Dirac's theory seriously* (Blackett 1962, p. 5, emphasis added).[26]

And when Dirac was asked if his proposal that an encounter of two gamma rays might lead to the creation of an electron and anti-electron bore "any close connection to Blackett's work," he answered:

I don't think so. I think *the experimental people go on their way, pretty well independently.* But you could ask Blackett about that (Dirac 1963, p. 4).

Thus, it's not surprising that Dirac's theory made only a belated appearance in Blackett and Occhialini's (1933) discovery report. And when it does, it comes

[26]See, for example: "In the beginning, these ideas seemed incredible and unnatural to everybody. No positive electron had been seen at that time; the asymmetry of charges ... seemed to be a basic property of matter" (Weisskopf 1983, p. 64); "[T]he Dirac theory, in spite of its successes, carried with it so many novel and seemingly unphysical ideas, such as negative mass, negative energy, infinite charge density, etc. Its highly esoteric character was apparently not in tune with most of the scientific thinking of the day. Furthermore, *positive electrons apparently were not needed to explain any other observations*." (Anderson 1961, p. 827, emphasis added).

under the heading "The Hypothetical Properties of the Positive Electron." In other words, once Blackett and Occhialini had experimentally established its existence and obtained a rough measure of its mass and charge, they then turned to Dirac and wondered what could be *hypothesized* about the life and death of the positive electron. Blackett and Occhialini began by noting that "[i]t is clear that [positive electrons] can have only a limited life as free particles since they do not appear to be associated with matter under normal circumstances (Blackett and Occhialini 1933b, p. 714)." And rather than enter into combination with other elementary particles "it seems more likely that they disappear by reacting with a negative electron to form two or more quanta (p. 714)." And it is at this point that Dirac's "hole" theory makes it appearance because it incorporates a "mechanism" for the short life span of the positron and its disappearance with only a few quanta left behind.

> On Dirac's theory the positive electrons should only have a short life, since it is easy for a negative electron to jump down into an unoccupied state, so filling up a hole and leading to the simultaneous annihilation of a positive and negative electron, the energy being radiated as two quanta (p. 714).

Moreover, this proposed mechanism can be shown to have predictive consequences, and Dirac in fact provided Blackett and Occhialini with a calculation of "the actual probability of this annihilation process" which is described in detail and yields the general prediction that "mean life is proportional to the concentration of negative electrons" which in the case of water yields the specific prediction of a "mean life of 3.6×10^{-10} s (pp. 715–716)." Thus, looking ahead, Blackett and Occhialini conclude that:

> When the behaviour of the positive electrons have been investigated in more detail, it will be possible to test these predictions of Dirac's theory. *There appears to be no evidence as yet against its validity*, and in its favour is the fact that it predicts a time of life for the positive electron that is long enough for it to be observed in the cloud chamber but short enough to explain why it had not been discovered by other methods (p. 716, emphasis added).

If all you had to go on were the textbook accounts discussed earlier, you would be surprised by Blackett and Occhialini's rather muted and qualified support for "Dirac's theory." And as well, we expect, by the lack of any kudos for Dirac's prediction of the positive electron in the first place. Instead they immediately moved on to produce a list of proposed topics for further investigation including whether "the anomalous absorption of gamma-radiation by heavy nuclei may be connected with the formation of positive electrons and the re-emitted radiation with their disappearance (p. 716, footnote omitted)."

3.5 Further Experimental Research Using Gamma Ray Radiation

Blackett and Occhialini's proposed research regarding the production of positrons by gamma ray interaction was undertaken and reported in Chadwick et al. (1933).

In addition, Blackett separately published a report in *Nature* reviewing his efforts, as well of those of Anderson among others, to "obtain further information as to the properties of positive electrons" by studying "in detail the simplest case known of their production; namely, that in which a beam of homogeneous gamma rays is absorbed by heavy elements (Blackett 1933, p. 917)." After presenting evidence indicating that "positive electrons are… produced outside the nucleus," Blackett assuming that this is so (and utilizing additional evidence at hand) went on to derive three additional conclusions (p. 917).

> (a) Since there is certainly no room, in atomic theory, for the permanent existence of positive electrons well outside a nucleus, then a positive electron that comes from there must be born there, and if born there, an equal negative electron must be born simultaneously in order to conserve electric charge.. ..[27]
>
> (b) The positive electron must have a spin of ½ and so obey the Fermi-Dirac statistics.. ..
>
> (c) A necessary consequence of the occurrence of the process whereby a quantum interacts with an atom to produce a pair of electrons of opposite sign, is the occurrence of the reverse process, in which a positive electron and a negative electron interact with each other and the field of an atom to produce a single quantum of radiation.. .. (Blackett 1933, pp. 917–918).

Regarding these conclusions, Blackett made a point of emphasizing that:

> These conclusions as to the existence and the properties of positive electrons *have been derived from the experimental data by the use of simple physical principles*. That Dirac's theory of the electron predicts the existence of particles with just these properties, gives strong reason to believe in the essential correctness of his theory (p. 918, emphasis added).

Blackett's procedure here should create in the reader a strong sense of déjà vu because it is virtually identical to that employed by Meselson who also made a point of isolating what his experimental data revealed on the basis of only basic assumptions about his experimental methodology. And like Blackett, Meselson took pains to distinguish those results from what was more deeply theoretical, in his case the Watson-Crick DNA hypothesis. By way of further elaboration, Blackett proceeded to spend several paragraphs expounding on the difference between the many successful predictions made by the Dirac Equation and its positive energy solutions *as opposed to* the "hole" theory which was introduced to make use of the negative solutions (see p. 918). Given this distinction, Blackett went on to state that:

> The experimental discovery of the positive electron has therefore removed a very serious theoretical difficulty, and by so doing, has greatly extended the field of phenomena over which Dirac's theory may be applied (p. 918).

---

[27]The conclusion that positrons and electrons are "born in pairs" as revealed in the "shower" photographs was first proposed in Blackett and Occhialini (1933, pp. 713–714) and then further pursued and here substantiated by the results obtained using gamma ray radiation instead of cosmic rays. Anderson for his part was generous in giving full credit to Blackett and Occhialini: "The idea that they were created out of the radiation itself did not occur to me at that time, and it was not until several months later when Blackett and Occhialini suggested the pair-creation hypothesis that this seemed the obvious answer to the production of positrons in the cosmic radiation. Blackett and Occhialini suggested the pair-production hypothesis in their paper published in the spring of 1933, in which they reported their beautiful experiments on cosmic rays using the first cloud chamber which was controlled by Geiger counters" (Anderson 1961, p. 826).

It should now be clear that Blackett's muted endorsement of "the Dirac theory" reflected his reservations (shared by many) about Dirac's incorporation of the negative energy solutions as an integral component of what became known as the Dirac Sea and its holes. The discovery of the positron served then to connect this otherwise fantastical but wonderfully imaginative construction to what was experimentally accessible and thus along with Dirac's computation of positron life expectancy served to add at least a patina of respectability to the negative energy side of the implications of the Dirac Equation.

There is, of course, much more to be said about the experimental and associated theoretical work that was undertaken in response to the discovery of the positron by Anderson, and Blackett and Occhialini. But for now it's clear that with respect to (Anderson 1933) and (Blackett and Occhialini 1933), once was enough to forgo any exact, or nearly so, duplication of those experiments, and to justify the pursuit of the avenues of further development that we have, if only briefly, mentioned here.

3.6 Summary and Conclusions

Once was clearly enough for the 1933 experiments of Anderson, and Blackett and Occhialini considered jointly to have established the existence of the positron. Although we would add that the results reported in Anderson (1933) were by themselves sufficient to have decisively done so. And if not for Anderson's results, we doubt that Blackett and Occhialini would have been held by many to have so clearly done so. This because of the inferior accuracy obtained due to the smaller magnetic field employed by Blackett and Occhialini and the consequential loss of measurement sensitivity.[28]

Once was enough for (Anderson 1933) and (Blackett and Occhialini 1933), either singly or in conjunction, to have justified the *pursuit* of experimental data and theoretical development relevant to the *production* of positrons by gamma rays including, for example, the pair-wise theory of positron creation proposed in Blackett and Occhialini (1933, pp. 713–714) and substantiated in Blackett (1933, pp. 917–918).[29]

1. Because of its many exotic and problematic features, once was not enough even for the combination of the 1933 results of Anderson, and Blackett and Occhialini to have confirmed Dirac's hole theory in anything other than the provisional and qualified sense elaborated in Blackett (1933). Still, this was enough to get Dirac over the finish line when it came to the award of the 1933 Nobel Prize in Physics. Although Dirac had received very few nominations, Anderson's discovery played a crucial role in the decision to make the award to Dirac. Committee member

[28] As an admittedly small piece of evidence on this score we note that Anderson received his Nobel Prize in 1936 while it took more than a decade of many more accomplishments before Blackett received his in 1948.

[29] For an extensive review of these later gamma rays experiments and the associated theoretical developments see Roque (1997).

Henning Pleijel in his presentation speech said to Dirac: "The experimental discovery of the existence of the positron has in a brilliant way confirmed your theory (quoted in Larsson and Balatsky 2019, p. 52)."

And while Dirac's hole theory, with the assistance of the experimentalists, may have thereby achieved an undeniable level of respectability, there were still certain remaining mathematical constructs that rankled many of his fellow theoreticians. In this regard see the following appraisal by Victor Weisskopf that brings to the fore, for example, the objectionable infinite charge densities.

> In spite of all the successes of the new hole theory of the positron, the infinite charge density and the infinite negative energy density of the [Dirac] vacuum made it very difficult to accept the theory at its face value. A war against infinities started at that time (Weisskopf 1983, 68).

Thus, as a coda to our conclusions we note that by 1934 the seeds were already being planted for what eventually would become quantum electrodynamics.[30]

References

Alvarez, L. 2014. Interview of Luis Alvarez, February 14, 1967. In *Niels Bohr library and archives*, ed. C. Weiner and B. Richman. College Park, MD: American Institute of Physics.

Anderson, C.D. 1931. Cosmic rays disrupt atomic hearts. *Science News Letter, (December 19, 1931)*, 387.

Anderson, C.D. 1932a. Energies of cosmic ray particles. *Physical Review* 41: 405–421.

Anderson, C.D. 1932b. The apparent existence of easily deflectable positives. *Science* 76: 238–239.

Anderson, C.D. 1933. Cosmic-ray bursts. *Physical Review* 43 (5):368–369.

Anderson, C.D. 1933a. The positive electron. *Physical Review* 43: 491–494.

Anderson, C.D. 1933b. Cosmic-ray positive and negative electrons. *Physical Review* 44: 406–416.

Anderson, C.D. 1934. The positron. *Nature* 133 (3357): 313–316.

Anderson, C.D. 1961. Early work on the positron and muon. *American Journal of Physics* 29: 825–830.

Anderson, C.D. 1965. The production and properties of positrons. *Nobel Lectures, Physics, 1922-1841*: 365–376 (Amsterdam, Elsevier).

Anderson, C.D. 1966. Interview of Carl Anderson, June 30, 1966. In *Niels Bohr library and archives*, ed. C. Weiner. College Park, MD: American Institute of Physics.

Anderson, C.D. 1979. Carl Anderson Interview, January 9-February 8, 1979. In *Archives California Institute of Technology*, ed. H. Lyle. Pasadena, California: California Institute of Technology.

Anderson, C.D., and H.L. Anderson. 1983. Unraveling the particle content of cosmic rays. In *The birth of particle physics*, ed. L. M. Brown and L. Hoddeson, 131–154. Cambridge: Cambridge University Press.

Blackett, P.M.S. 1933. The positive electron. *Nature* 132 (3346): 917–919.

Blackett, P.M.S. 1962. Interview of P. M. S. Blackett, December 17,1962. In *Niels Bohr library and archives*. ed. J. L. Heilbron. College Park, MD: American Institute of Physics.

Blackett, P.M.S., and G.P.S. Occhialini. 1933. Some photographs of the tracks of penetrating radiation. *Proceedings of the Royal Society (London)* A139: 699–727.

Cahn, R., and G. Goldhaber. 2009. *The experimental foundations of particle physics*. Cambridge: Cambridge University Press.

[30]For brief but incisive reviews of this "war against infinities" and the development of quantum electrodynamics see Weisskopf (1983, pp. 68–78) and Pashby (2012, pp. 455–457).

Chadwick, J., P.M.S. Blackett, et al. 1933. New evidence for the positive electron. *Nature* 131 (3309): 473.

De Maria, M., and A. Russo. 1985. The discovery of the positron. *Rivista di storia della scienza* 2: 237–286.

Dirac, P.A.M. 1928a. The quantum theory of the electron. *Proceedings of the Royal Society (London)* A117: 610–624.

Dirac, P.A.M. 1928b. The quantum theory of the electron, Part II. *Proceedings of the Royal Society (London)* A118: 357–361.

Dirac, P.A.M. 1930. On the annihilation of electrons and protons. *Proceedings of the Cambridge Philosophical Society* 26: 361–375.

Dirac, P.A.M. 1931. Quantised singularities in the electromagnetic field. *Proceedings of the Royal Society (London)* 133 (821): 60–72.

Dirac, P.A.M. 1963. Interview of P. A. M. Dirac May 14, 1963. In *Niels Bohr library and archives*, ed. T. Kuhn. College Park, MD: American Institute of Physics.

Dirac, P.A.M. 1983. The origin of quantum field theory. In *The birth of particle physics*, ed. L.M. Brown and L. Hoddeson, 39–55. Cambridge University Press.

Galison, P. 1997. *Image and logic*. Chicago: University of Chicago Press.

Griffiths, D. 2008. *Introduction to elementary particles*. Weinheim: Wiley.

Larsson, M., and A. Balatsky. 2019. Paul Dirac and the nobel prize. *Physics Today* 72: 46–52.

Martin, B.R. 2009. *Nuclear and particle physics*. Chichester: Wiley.

Millikan, R.A. 1935. *Electrons (+ and -), protons, photons, neutrons, and cosmic rays*. Chicago: University of Chicago Press.

Millikan, R.A., and C.D. Anderson. 1932. Cosmic-ray energies and their bearing on the photon and neutron hypotheses. *Physical Review* 40 (3): 325–328.

Oppenheimer, J.R. 1930. On the theory of electrons and protons. *Physical Review* 35 (5): 562–563.

Pashby, T. 2012. Dirac's prediction of the positron: A case study for the current realism debate. *Perspectives on Science* 20 (4): 440–475.

Roque, X. 1997. The manufacture of the positron. *Studies in History and Philosophy of Modern Physics* 28: 73–129.

Skobeltzyn, D. (1983). The early stage of cosmic-ray particle research. In *The birth of particle physics*, ed. L.M. Brown and L. Hoddesn, 111–119. Cambridge University Press.

Staley, K. 1999. Golden events and statistics: What's wrong with Galison's image/logic distinction. *Perspectives on Science* 7: 196–230.

Weisskopf, V.F. (1983). Growing up with field theory: The development of quantum electrodynamics. In The birth of particle physics, ed. L.M. Brown and L. Hoddeson, 56–81. Cambridge University Press.

Weyl, H. 1932. *The theory of groups and quantum mechanics (Second revised edition)*. Robertson, H.P. 2014. *Gruppentheorie und Quantenmechanik*. Mansfield Centre: CT, Martino Publishing.

Chapter 4
The Discovery of the Omega Minus Hyperon

This section deals with the second of the purportedly "golden event," discoveries that we promised to examine, namely, that of Omega Minus hyperon. As will be seen, its discovery had two significant benefits. First, it solidified a promising method of imposing order on the 60 or so experimentally confirmed candidates for the status of being distinct sub-atomic particles that existed at the time.[1] And second, it facilitated the subsequent development of a deeper theory of the sub-atomic world that involved the introduction of quarks as fundamental constituents.

4.1 The Eightfold Way

One strategy for imposing order on what was this otherwise unruly bunch of particle candidates was to embed them in the confines of group theory, and more particularly, into what are known as Lie groups (where the group elements are continuous functions). The most successful such coopting of group theory in this period was what became known as the Eightfold Way, which was independently developed by Gell-Mann (1961) and by Ne'eman (1961).[2] The basis for the approach was to take a certain class of symmetries (i.e. conservation and translational relations) as fundamental but where those symmetries could be broken or perturbed.[3]

[1]For an overview of the particle candidates and the various attempts to impose order see Chew et al. (1964). For a comprehensive enumeration of the denizens of the particle zoo for this period see Gell-Mann and Rosenfeld (1957).

[2]For an elegant and comprehensible introduction to the Eightfold Way see (Ne'eman and Kirsh 1986, pp. 198–205). Copies of the major papers are conveniently collected in Gell-Mann and Ne'eman (2018) which also includes commentary by Murray Gell-Mann and Yuval Ne'eman.

[3]The innovation here was to deliberately use first order approximations in calculating the interaction of some (but not all) of the symmetries where the approximate nature of the calculations allowed for the symmetries to be broken in consequential and what turned out to be empirically substantiated ways. See Gell-Mann (1962a) and Gell-Mann and Ne'eman (2018, pp. 216, 338, 369).

A. Franklin and R. Laymon, *Once Can Be Enough*,
https://doi.org/10.1007/978-3-030-62565-8_4

Initially, it was shown that the Eightfold Way entailed a coherent "octet" form of organization that could be filled by either the semi-stable mesons (spin 0), the semi-stable baryons (spin ½), or the mesonic-resonances (spin 1). (See, for example, Figs. 4.1 and 4.2, for the octet of semi-stable mesons (spin 0) and the octet of the semi-stable baryons.) At the time each of these groups were missing at least one

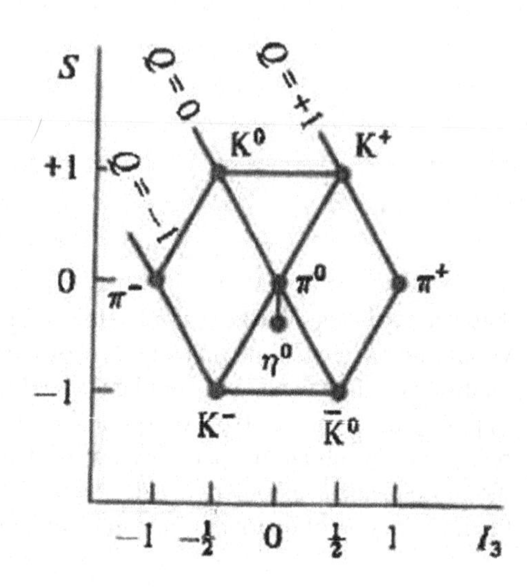

Fig. 4.1 The Meson Octet

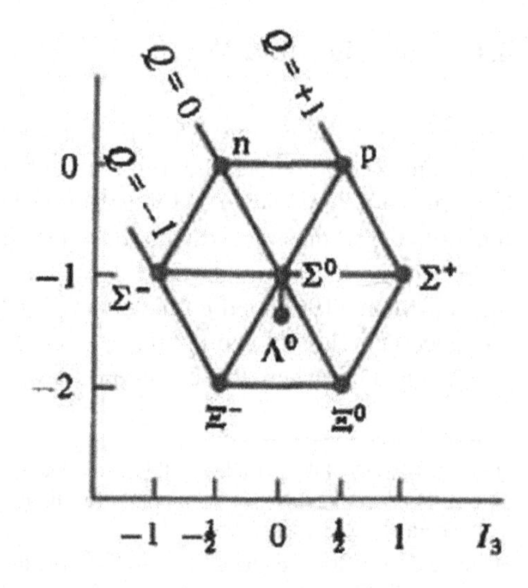

Fig. 4.2 The Baryon Octet

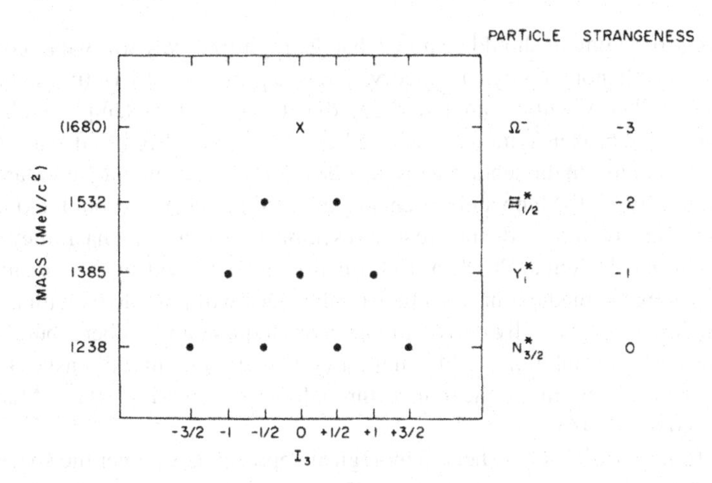

Fig. 4.3 The Baryon Decuplet. *Source* Barnes et al. (1964a)

particle.[4] Nevertheless, the model did have empirical support because many of the spaces in the groups were filled with known particles.

Not everyone, at this time, was entranced by this success. Harry Lipkin in his comprehensive review of the development of the quark model made the following appraisal of the situation before the certified arrival of Ω^-.

> At that time the eightfold way was generally considered to be rather far-fetched and probably wrong. . . . Several years later, new experimental data forced everyone to take SU(3) [the underlying group structure employed in the Eightfold Way] more seriously. The second baryon multiplet was found . . . including the Ω^- . . . (Lipkin 1973).

The octet for the semi-stable baryons (spin ½) was especially noteworthy because it predicted a spin of ½ for the xi particles (Ξ^0, Ξ^-), while the principal competitor model predicted 3/2 as the spin.[5] The matter was decided in 1963 when the spin was experimentally determined to be ½.[6] The most striking success of the Eightfold Way, however, was to be found in the decuplet structure imposed by the theory on the baryons of spin 3/2. At the time the predicted inhabitants of the decuplet at the mass positions 1530 and 1672 were as yet undetected. See Fig. 4.3.

[4]In Figs. 4.1 and 4.2 the particles are plotted against strangeness (S), see note 11, and I_3, the third component of isotopic spin or isospin. Isospin is a quantum number that is conserved in the strong interactions. Isospin values are calculated using the number of particles in a multiplet, subtracting one, and dividing by two. Thus, in Fig. 4.1 there are three pions ($\pi^+\pi^o\pi^-$) so the isospin of the pion is 1. I_3 is given by $I_3 = Q - (S + B)/2$. Where Q is the charge of the particle in units of the electron charge, S is its strangeness, and B is its baryon number. For the π^o, $Q = S = B = 0$ so $I_3 = 0$. The other particle at the center of the octet, the η, is an isospin singlet with $I = I_3 = 0$. It is a different particle from the π^o with a different mass and a different lifetime.

[5]We will discuss the particle missing from the octet of semi-stable mesons below.

[6]See Ne'eman and Kirsh (1986, p. 202), Gell-Mann and Ne'eman (2018, p. 368), and Dothan et al. (1962).

Before proceeding, it should be noted that the Eightfold Way was what physicists describe as a "phenomenological" theory, that is, a primarily descriptive and formal mathematical theory without an underlying dynamics. So, for example, Gell-Mann noted that "[t]he broken symmetry picture [contained in the Eigthfold Way] is hard to interpret on any fundamental theoretical basis" (Gell-Mann and Ne'eman 2018, p. 87). Similarly, Gell-Mann and Ne'eman explained that they "set out looking for a formalism that would provide the greatest resemblance to phenomena as they stood" (Gell-Mann and Ne'eman 2018, p. 87). Finally, as expressed by Ne'eman: "One really hopes that someday our dynamicist colleagues will provide us with a theory from which we shall trivially derive all our internal quantum numbers; but this still looks like wishful thinking. . . . My suggestion is that no harm will ensue if we do solve some problems 'in the meantime' through our symmetries." (Gell-Mann and Ne'eman 2018, p. 314).

One problem with such a phenomenological approach is whether the specifics of the formalism represent an underlying reality or are just some sort of mathematical artifact.[7] In particular, the problem posed for the Eightfold Way was whether the remaining slots in the decuplet reflected real possibilities or were just a quirk of the formalism.[8]

But things changed when at the 1962 particle physics conference at CERN, it was announced that the Ξ* resonance at 1535 MeV was detected.[9] "The finding of the Ξ(1530) at the CERN '62 conference had an enormous consequence (Samios 1993, p. 9)." Gell-Mann who was at the conference immediately realized the consequences for the decuplet, and in a discussion section explained where the Ξ* resonance, the Ξ(1530), fit into the decuplet and that only one slot remained to be filled at the 1685 MeV level. He then dubbed the yet to be discovered particle the Omega Minus (Ω^-).[10] In other words, that remaining available slot could readily be interpreted as a

[7] A helpful analogy here is with Ptolemaic astronomy which while it had accurate orbital calculations (with respect to an earth centered celestial sphere), it did not provide a genuine dynamic explanation. The epicycles thus were purely a mathematical convenience, which Copernicus also retained. A fully dynamic account complete with elliptical orbits was not achieved until Newton.

[8] The worry about whether the consequences of the Eightfold Way were mathematical artifacts was for some practitioners more than just an idle concern as exemplified in Oakes and Yang (1963) who concluded that "if the Ω^- is found and if it does satisfy the equal-spacing rule, it can hardly be interpreted as giving support to the octet symmetry model, at least not without the introduction of drastically new physical principles." But such concerns were put to rest by, among others, (Eden and Taylor 1963; Frautschi 1963). For a review of the debate see Gell-Mann and Ne'eman (2018, pp. 369–70). George Zweig, who along with Gell-Mann first proposed the quark theory, admitted that "I didn't understand the Oakes-Yang paper and hoped that the problems they raised would somehow go away. Eventually they did (Zweig 1980)."

[9] Bertanze et al. (1962) and Pjerrou (1962).

[10] For more details about this episode see (Gell-Mann 1962b), (Gellman and Ne'eman 2018, pp. 340, 368–69).

prediction of a yet unknown particle with mass 1685 MeV, strangeness −3 and spin 3/2.[11]

4.2 The Omega Minus Hyperon

Given the status of the Eightfold Way as essentially a phenomenological theory and the suspicion that certain consequences might be merely mathematical artifacts, the experimental verification of the existence of Ω^- was understandably taken to be "a crucial test of the broken symmetry" that was the central component of the Eightfold Way (Gell-Mann and Ne'eman 2018, p. 86). Thus, it is not surprising that after making the Ω^- prediction, Gell-Mann contacted Maurice Goldhaber, the Director of the Brookhaven National Laboratory, and requested that he initiate a search for Ω^-. (Gellman and Ne'eman 2018, p. 340). Goldhaber obliged and after a two-year effort the Ω^- was found.

Because there were no strongly interacting particles whose strangeness added up to −3 and which also had a total mass less than that of the proposed Ω^- that particle would have to decay through the weak interaction with a lifetime long enough so that it would leave a noticeable track in the bubble chamber.[12]

> Consideration of the various conservation rules narrowed down the ways in which the omega-minus might decay into three most likely possibilities: it might break down into (1) a xi-minus particle and a neutral pi meson, (2) a xi-zero particle and a pi-minus meson or (3) a lambda-zero particle and a K-minus meson. These, then, were the results watched for in analyzing the photographs. (Fowler and Samios 1964, p. 40).[13]

Each of these decay modes would result in a Λ hyperon either in the initial decay of the Ω^- or in the subsequent decay of the xi particles. These would be indicated by a V in the photograph (see Fig. 4.4, tracks 5 and 6). In this event it was shown that the Λ could not come from the initial production vertex and therefore must be a decay product. The proposed production and decay processes for the initial Ω^- are shown in Fig. 4.5.

One feature of the experimental analysis that needs to be emphasized is there was an initial production and subsequent decay process where both were subjected to critical analysis and shown consistent with theoretical expectations. So, for example, the following argument was given for the particle identification and decay process: "…we have used the calculated π^0 momentum and angles, and the values from

---

[11] Strangeness was property assigned to elementary particles in order to account for various oddities in the production and decay of some particles. It is conserved in the strong and electromagnetic interactions, but not in the weak interactions.

[12] As noted by Gell-Mann, "Ω^- would have the striking property of metastablity, decaying only by weak interactions and yielding $\pi^0 + \Xi^-$, $\pi^- + \Xi^0$, and K-+ Λ" (Gellman and Ne'eman 2018, p. 86).

[13] The branching ratios for the respective decays were $K^- + \Lambda$ (67.8 ± 0.7) %, $\pi^- + \Xi^0$ (23.6 ± 0.7) %, and $\pi^- + \Xi^0$ (8.6 ± 0.4) %.

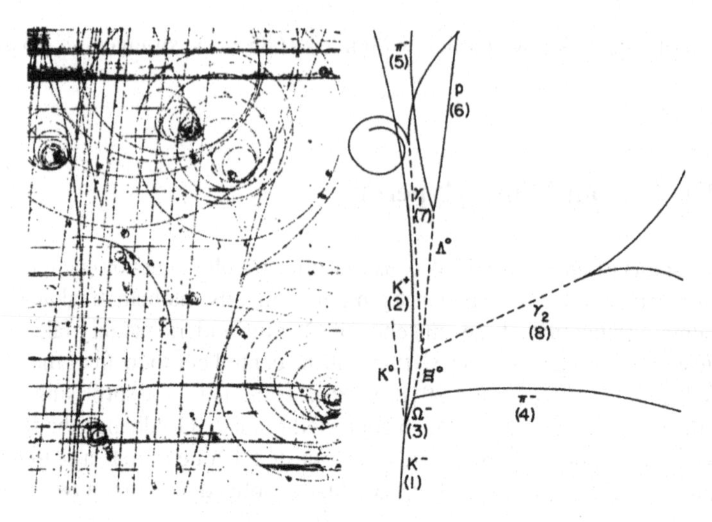

Fig. 4.4 The bubble chamber photograph showing the first Ω^- along with a schematic sketch of the event. *Source* Barnes et al. (1964a)

Fig. 4.5 The proposed production and decay processes for the initial Ω^-. *Source* Barnes et al. (1964a)

$$K^- + p \to \Omega^- + K^+ + K^0$$
$$\Omega^- \to \Xi^0 + \pi^-$$
$$\Xi^0 \to \Lambda^0 + \pi^0$$
$$\pi^0 \to \gamma_1 + \gamma_2$$
$$\gamma_2 \to e^+ + e^-$$
$$\gamma_1 \to e^+ + e^-$$
$$\Lambda^0 \to \pi^- + p.$$

the fitted Λ^0 to determine the mass of the neutral decaying hyperon to be 1316 ± 4 MeV/c^2 in excellent agreement with that of the Ξ^0 (Barnes et al. 1964a, pp. 204–205)."[14] The experimenters also noted that the measured transverse momentum of particle 4 in the diagram was 248 ± 5 MeV/c "greater than the maximum momentum for the possible decay modes of the known particles except for $\Xi^- \to e^- + n + \nu$.... We reject this hypothesis not only because it involves $\Delta S = 2$,[15] but also because it

[14]This was an internal calibration, demonstrating that the experiment could measure the mass of known particles well.

[15]Previous studies of weak interactions had shown that weak decays involve only a change in strangeness of +1.

disregards the previously established associations of the Λ and two gammas with the event (p. 205)." They further remarked that the lifetime of particle 3 (the proposed Ω^-), 0.74×10^{-10} s, indicated that it decayed by the weak interaction into a system with strangeness −2 and concluded that the Ω^- particle has strangeness minus 3.

They further noted that, "From the momentum and gap length measurements, track 2 is identified as a K^+ (A bubble density of 1.9 times minimum was expected for this track while the measured value was 1.7 ± 0.2). Tracks 5 and 6 are in good agreement with the decay of a Λ, but the Λ cannot come from the primary interaction. The Λ mass as calculated from the measured proton and π^- kinematic quantities is 1116 ± 2 MeV/c. Since the bubble density from gap length measurement of track 6 is $1.52 + 0.17$, compared to 1.0 expected for a π^+ and 1.4 for a proton, the interpretation of the V as a K is unlikely (p. 204)."

The final reported result then was that: "In view of the properties of charge ($Q = -1$), strangeness -3, and mass ($M = 1686 \pm 12$ Mev/c^2) established for particle 3, we feel justified in identifying it with the sought for Ω^- (p. 206)." Significantly, there was no explicit statistical argument offered for this result. Thus, if the decay path was indeed a "golden event," it was one without an explicit statistical argument.[16]

But this is not the end of the story because a few weeks later another decay mode of the Ω^- was identified, but this time it was the second of the three possibilities described above, namely, the decay into $K^- + \Lambda^0$. This discovery was separately reported with the title "Confirmation of the Existence of the Ω^- Hyperon."

> To summarize the above results, we believe that we have found a second example of the production and decay of an Ω^-, in this case the Ω^- decaying into a Λ^0 and K^-. The measured mass of 1674 ± 3 MeV/*c* is in excellent agreement with the previous determination of the Ω^- mass as well as the theoretically predicted value (1676 MeV/*c*). All other possible interpretations of the event have been eliminated on the basis of ionization and kinematic information (Barnes, Connolly et al. 1964b, p. 136).

Despite their obvious confirmational importance for the Eightfold Way, there were no calls for either of these experiments to be repeated. In short, the two reports were sufficient to establish the existence of a particle with charge −1, strangeness −3, and mass somewhere around 1680 MeV.

Thus, what remained to be shown was that this particle, tentatively identified as Ω^-, was in fact the particle predicted by Gell-Mann and Ne'eman on the basis of the Eightfold Way. As summarized by Fowler and Samios:

[16]We might suggest here that there was an implicit statistical argument offered namely that the production and decay processes offered were very unlikely on the basis of background processes. They were unique to the decay of a particle with negative charge and strangeness −3. The observed event itself was unusual. "The event is unusual in that two gamma rays, apparently associated with it, convert to electron-positron pairs in the liquid hydrogen (Barnes et al. 1964a, pp. 204–205, p. 204)." The probability of both gamma rays materializing into electron-positron pairs in the liquid hydrogen was 10^{-3}. There was an additional oddity. Nicholas Samios, one of the discoverers of the Ω^- later remarked, "It is also true that the γ conversions were not observed during the original discovery of this event; the scanner, who happened to be me, missed them. However, they were found the next day while the event was reexamined (Samios 1997, p. 534)."

> To verify the predictions of the eightfold way and establish the identity of the omega-minus particle beyond any question, three more proofs are wanted. The two photographs showed two different modes of decay of the omega-minus; we would like to see the third one that was predicted as being likely, namely, its decay into a xi-minus particle and a pi-zero meson. Then we need determinations of the spin and parity of the supposed omega-minus particle. *The two photographs establish its mass and strangeness quantum number satisfactorily, but for calculation of its spin and parity we shall need many more photographs of the event, because these properties can only be deduced from the statistical distributions of the angles involved in the various decays* (Fowler and Samios 1964, p. 45, emphasis added).

Thus, the agenda was set for the experimental *pursuit* of further information about Ω^-. The discovery of the third decay path was not long in coming. In November 1964 a third Ω^- event was reported illustrating the last of the three significant decay modes, $\Omega^- \rightarrow \Xi^- + \pi^0$ (Abrams et al. 1964). By the end of 1964 the Ω^- and its existence had received the imprimatur of the Particle Data Group (Rosenfeld et al. 1964).[17]

For the gradual acquisition of additional decay data and refinements in the determination of mass and lifetime see (Hemingway et al. 1978). Attempts at the spin determination of the Ω^- were not published until 1978. Three independent experimental groups announced slightly different conclusions, but were unable to definitively establish that the spin was 3/2. Hemingway and collaborators stated, "We have searched without success for reference frames which show sufficient alignment of the Ω^- to allow an independent spin determination. The probability of a flat distribution [spin ½] was always several percent or more. We are therefore unable to exclude $J = ½$ for the Ω^-. At 10 and 16 GeV/c the probability that the distribution is flat is less than 0.001 whereas the probability for spin 3/2 is about 0.7 (pp. 215–216)." Deutschmann and collaborators were more definitive. "A total of 101 Ω^- decays have been found in K^-p interactions at 10 and 16 GeV/c. The decay angular distributions have been fitted under the assumptions that the Ω^- has spin either ½ or 3/2. It has been found that the probability of isotropy (spin 1/2) is less than one in a thousand, whereas the probability for spin 3/2 is about 70% (Deutschmann et al. 1978, p. 96)." More cautiously Baubillier et al. concluded that, "the probability of consistency with a flat distribution ~1/300 indicating $J \neq ½$ (Baubillier et al. 1978, p. 384)."[18]

In 2006 the BaBar collaboration (Aubert et al. 2006) reported a quite different experiment that measured the spin of the Ω^-. That remarked that, "In previous attempts [they referenced the three 1978 papers discussed earlier] to confirm the spin of the Ω^-, K^-p interactions in a hydrogen bubble chamber were studied. In each case only a small Ω^- data sample was obtained and the Ω^- production mechanism was not well understood. As a result, these experiments succeeded only in establishing that the Ω^- spin is greater than 1/2 (p. 112001–4)." The group presented "A measurement

[17]The reports of the Particle Data Group indicate what is accepted knowledge by the particle physics community.

[18]Despite the successful prediction of Ω^- with its "striking" confirmed properties, there had to have been some apprehension about whether the spin prediction of 3/2 would also garner experimental confirmation. Richard Feynman taking advantage of this, came into Ne'eman's office and "seemingly excited," wickedly announced "Have you heard the news? The Omega Minus has a spin of one half! (Gell-Mann and Ne'eman 2018, p. 370)."

of the spin of the Ω^- hyperon produced through the exclusive process $\Xi_c^0 \to \Omega^- K^+$ is presented.... *Under the assumption that the Ξ_c^0 has spin ½* the angular distribution for $\Omega^- \to \Lambda^-$ decay is inconsistent with all half-integer Ω^- spin values other than 3/2. Lower statistics data for the process $\Omega_c^0 \to \Omega^- \pi^+ \ldots$ are also found to be consistent with Ω^- spin 3/2. If the Ξ_c^0 spin were 3/2 an Ω^- spin of 5/2 could not be excluded (112001–4)."

Once the spin of the Ω^- had been determined to be 3/2 one could determine the parity of the Ω^-. For a two-body final state, such as $K^-\Lambda$, the parity is given by (the intrinsic parity of particle 1, the K^-) × (the intrinsic parity of particle 2, the Λ) × $(-1)^l$, where l is the orbital angular momentum of the two final-state particles. The K^- has negative intrinsic parity and the Λ has positive intrinsic parity. The initial state, the Ω^- has spin 3/2. By conservation of angular momentum, the final state must have total angular momentum 3/2. The K^- has spin 0 and the Λ has spin ½. Thus, the final state must have orbital angular momentum 1. The parity of the final state is thus $(-1) \times (+1) \times (-1)^1 = +1$, and the parity of the Ω^- is therefore positive, as predicted by the Eightfold Way.

4.3 Summary and Conclusion

Question: Where does replication fit into this picture? While the results of the follow-up experiments described are not exact or even nearly so replications of (Barnes et al. 1964a), that sequence, beginning with (Barnes et al. 1964b), can readily be interpreted as a sequence of *implicit replications*. This because inconsistent or incoherent results would have cast doubt on the decuplet representation of the Eightfold Way as well as, though less directly, on (Barnes et al. 1964a). But since there was no such incoherence (and since the determined value of spin was 3/2 as predicted), there was accordingly this *cumulative* evidence that *if* the Brookhaven experiments were repeated, Ω^- would again be discovered.

The 1978 experimental determinations of spin present a revealing contrast when compared to the experimental proof for the existence Ω^- provided in Barnes et al. (1964a, 1964b). As already noted, the analyses of the decay paths in the Brookhaven experiments were by and large devoid of statistical argumentation. On the other hand, as can be readily confirmed by even a cursory examination, the supporting analyses for the spin determinations are thick with statistical argumentation and accordingly came with associated measures of statistical uncertainty. Importantly, the measures of uncertainty in the 1978 reports provided a *target* for the reduction of statistical uncertainty and hence *a reason for further replication* in the sense of obtaining less uncertain values.[19] Thus, in 2006, the stronger (but carefully conditional) conclusion was that:

[19] We've extensively examined this sort of replication with respect to null experiments in Franklin and Laymon (2019).

Under the assumption that the Ξ_c^0 has spin 1/2, the angular distribution of the Λ from $\Omega^- \rightarrow \Lambda K^-$ decay is *inconsistent with all* half-integer Ω^- spin values other than 3/2 (Aubert et al. 2006, pp. 2–3, emphasis added).[20]

In sum, once was enough for the combination of (Barnes et al. 1964a, 1964b) to establish the *existence* of Ω^- *and* to justify the effort to experimentally gather increasingly more accurate and certain information about the particle.

There may, however, have been more to the decision that once was enough than just the obvious quality of the photographic evidence and analysis of (Barnes et al. 1964a, 1964b)[21]—at least for *some* of the theoretical and experimental practitioners. What caught our attention on this point was the following feature of the Brookhaven experiment, namely, that Ω^- had in fact *not been detected* in the initial 50,000 bubble chamber photographs. But despite this lack of success, the search continued for nearly another 50,000 photographs before the confirming events were found (Gell-Mann and Ne'eman 2018, p. 370). All told, nearly two million feet of K^- track length had been scanned in the search. This suggests that the earlier empirical success of the Eightfold Way weighed in favor of replication (i.e., continuing the experimental search) after the initial failure to detect.[22]

Such a weighing in favor of replication on the basis of past theoretical success is just an application of the truism that if the prediction of a *previously successful* theory fails, that is a reason for suspicion of the result and replication of the experiment. But if an experiment is found confirming, the other side of the truism applies, namely, when a prediction succeeds, that is a reason not to replicate the experiment. In other words, when an experimental result *is expected*, a theory can be said to confirm the correctness of experimental procedure and execution even when the outcome of the experiment serves to confirm the theory. As mentioned in our introductory section, these relations constitute an important element in what we have described as an *epistemology of experiment*. There is, of course, always in all of this the possibility of otherwise overriding considerations.[23]

---

[20] But see Aubert et al. (2006, p. 7) for a more completely stated description of the experimental result and its conditional nature.

[21] To this must be added the reputational value of the Brookhaven National Laboratory.

[22] The earlier empirical success of the Eightfold Way also must have weighed heavily in favor of initiating the Brookhaven search for Ω^- in the first place because, as noted by Samios, "the competition for running time and resources was ferocious (1994, p.4)."

[23] For an example of these sorts of interaction between theory and experiment consider Gell-Mann's answer to an interviewer's question: "How much do you fight for your theories if it looks as if they've been proved wrong?" Answer: ". . . if it's a very complicated experimental situation; the theory looks particularly beautiful; you might hope there's something the matter with the experiment. They're very difficult in this field; they take a long time, they're very expensive, and they're very hard to recheck. So that it quite often happens that an experimental result that's important is really not right (Gell-mann)." In a similar vein, Samios raised the question of "why [the] discovery of baryon charm was not immediately embraced by the high energy community. An obvious answer is that it was only one event. But the Ω^- discovery consisted of one event, why the difference? Unlike SU(3) the theoretical underpinning for charm was less firm. There was no spectroscopy and the mass predictions for charm particles were not precise. Probably more important was the negative experimental findings by the MARK II detector at SPEAR/SLAC." In other words, because

With the above in mind, consider the following interaction between Ne'eman and the Brookhaven experimentalists:

> Nick Samios and the other experimentalists . . . forgot to let me know about the [Ω^-] results . . . . I sent in a request for a reprint, and received a nice set of bubble chamber pictures, with the note "Please excuse the over-sight, but you knew it existed before we did!" (Gellman and Ne'eman 2018, p. 370).

So for Ne'eman, since he already "knew" (on the basis of the Eightfold Way and its earlier success) that Ω^- exists, and since the Brookhaven experiment confirmed what Ne'eman already "knew," there was (in the absence of otherwise overriding considerations) no reason for Ne'eman to doubt the experiment or think that replication was called for. While offered somewhat tongue in check, we think the episode nevertheless revealing as to the more complicated picture proposed here of the confirmational relationship between theory and evidence.

Therefore, skeptics of the Eightfold Way would not have viewed its purported successes as "confirming" (in the sense described above) the correctness of the experimental procedure and execution employed in Barnes et al. (1964a). But as Lipkin's appraisal, discussed above, indicates, the skeptics eventually were "forced" into accepting the discovery of Ω^- as cumulatively confirmed by the "new experimental data," and as a consequence to "take SU(3) more seriously."

In sum, for both enthusiast and skeptics, there emerged agreement that the discovery of Ω^- justified the *pursuit* and *development* on the basis of the Eightfold Way of a physics that promised to be revealing of the so far missing underlying dynamics. And the results of that pursuit were not long in coming since in 1964 Gell-Mann and George Zweig independently made use of the Eightfold Way as their starting point for their initial versions of the quark theory.[24]

Interestingly, Haim Goldberg-Ophir and Ne'eman (1962) had before the discovery of Ω^- developed a precursor theory of quarks similarly based on the Eightfold Way but where:

> The theory did not attract much attention, both because the eightfold way had not yet won general recognition (the Ω^- had not yet been discovered) and also because it did not go far enough. The authors had developed the mathematics resulting from the eightfold way, but they had not yet decided whether to regard the fundamental components as proper particles or as abstract fields that did not materialize as particles (Ne'eman and Kirsh 1986, p. 206).[25]

So evidently, the discovery of Ω^- played an important role in the ability of theoretical physicists to *attract attention* to their work—which raises fascinating questions

---

of the absence of firm theoretical support, the negative experimental findings were accepted, at least initially, at face value and replication was not in order. Though ultimately the situation was "subsequently rectified." (Samios 1993, p. 20) As we've noted, there's always the possibility of overriding considerations including different appraisals of the risk involved in the pursuit of new theories.

[24]Gell-Mann (1964) and Zweig (1964a, 1964b).

[25]For a more thorough discussion of Goldberg and Ne'eman (1962) and its reception see (Gell-Mann and Ne'eman 2018, pp. 366–368). See also Lipkin (1973, p. 180) for an analysis of why the Ne'eman-Goldberg proposal was not taken seriously before the discovery of Ω^-.

of what would have happened if the situation had been otherwise and say Brookhaven had abandoned the search for Ω^- after the failure to discover it in the initial set of 50,000 collision photographs.[26]

Further expansion, however, on the transition to the quark theory must await another time. For now we rest our case that with respect to the discovery of Ω^-, once was enough (1) to avoid exact (or nearly exact) replication; (2) to establish the existence of Ω^-; (3) to justify further experimentation on the spin and other properties of Ω^-; and (4) to justify the pursuit on the basis of the Eightfold Way of a physics that was more revealing of the underlying dynamics.

An interesting sidelight to this episode concerns the discovery of the η meson by Pevsner and collaborators (1961). The background to this experiment was both experimental and theoretical. The experimenters remarked that, "Many authors have speculated on the existence of neutral, strongly interacting bosons of mass of the order of 3–4 m_π, in order to fit the data for nucleon form factors obtained from electron scattering experiments (Pevsner et al. 1961, p. 421)."

The Pevsner experiment was conducted using photographs obtained in the Berkeley 72-inch bubble chamber exposed to a 1.23 Gev/c positive pion beam from the Berkeley Bevatron. The group remarked that the bosons that had been theoretically suggested "could be readily observed in the reaction $\pi^+ + d \rightarrow p + p + X^0$." To observe the possible decay $X^0 \rightarrow \pi^+ + \pi^- + \pi^0$ they examined the reaction $\pi^+ + d \rightarrow p + p + \pi^+ + \pi^- + \pi^0$. The results are shown in Fig. 4.6. The large peak at approximately 770 MeV was the then recently discovered ω^o meson. The smaller peak at 550 MeV was the η meson. No mention was made of the Eightfold Way or the fact that this particle might fill the one remaining place in the pseudoscalar meson octet (see Fig. 4.1). That identification was not made until Bastien et al. in early 1962. "We conclude that our results are most consistent with the quantum numbers 0^{-+} for the η (these are the same as the quantum numbers of the χ meson introduced in the "eightfold way" of Gell-Mann[9]). Statistical limitations and background do not permit us to rule out the case 1^- with certainty (Bastien et al. 1962, p. 117)."[27]

The initial discovery by Pevsner's group or even the identification of the η as a particle predicted by the Eightfold Way by Bastien et al. did not attract as much attention as the discovery of the Ω^- hyperon and was not regarded as strong support for the Eightfold Way. This was despite the fact that, in their initial publications, both Gell-Mann and Ne'eman had predicted such a particle. "The most clear-cut new prediction for the pseudoscalar mesons is the existence of χ^o, which should decay into 2γ like the π^0, unless it is heavy enough to yield $\pi^+ + \pi^- + \gamma$ with appreciable probability…. $\chi^o \rightarrow 3\pi$ is forbidden… (Gell-Mann 1961) reprinted in Gell-Mann and Ne'eman (2018, p. 41)."[28] Ne'eman predicted a similar particle, which he called the $\pi^{0'}$ to distinguish it from the π^0.

---

[26]If more photographs had already been taken this was unlikely.

[27]The experimental group acknowledged communications with and assistance of Gell-Mann.

[28]Gell-Mann was mistaken. As we have seen the initial discovery of the η meson was in a three-pion state

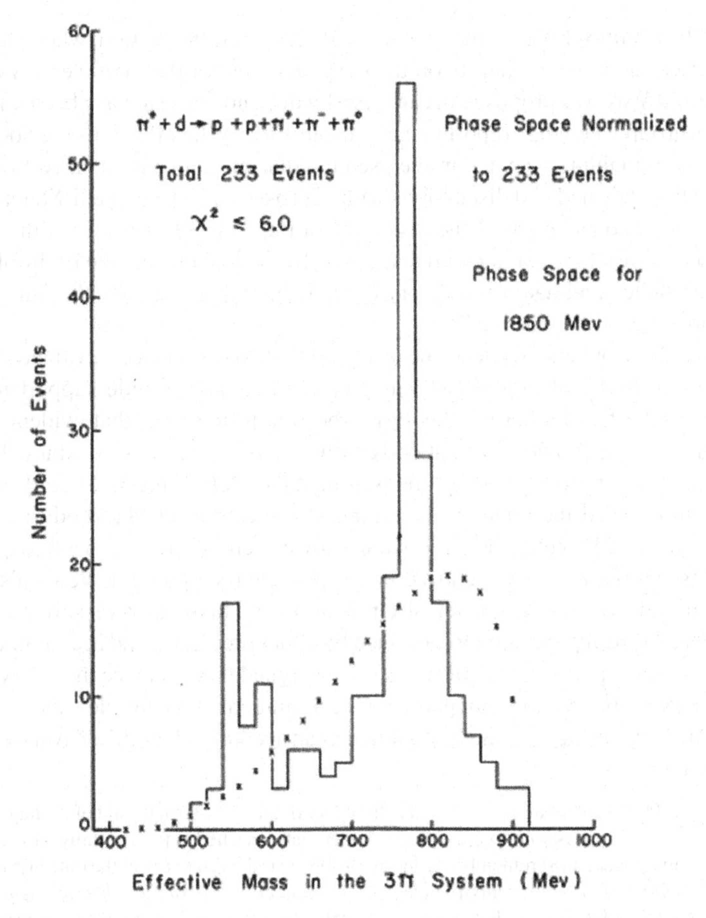

Fig. 4.6 Histogram of the effective mass of the three-pion system for 233 events. *Source* Pevsner et al. (1961)

As Arthur Rosenfeld remarked, "The η(548) meson had been predicted by Gell-Mann's Eightfold Way (3) in 1961. It was expected to decay into two photons or four pions. A meson (apparently unrelated because it decayed into three pions was discovered late in 1961, and properly identified as the predicted pseudoscalar meson, the η, early in 1962. This completed the first meson octet, but by later standards it attracted little attention (no press conference, no flurry of theoretical papers, and no 1962 edition of UCRL-8030) (Rosenfeld 1975, p. 559)."[29]

Why was the discovery of the Ω^- regarded as a strong confirmation of the Eightfold Way, whereas the discovery of the η wasn't. They were both predicted by the theory and both particles filled a sole remaining place in one the particle groups. One may speculate here that the difference was due to the difference in the evidential

[29]The UCRL report referred to was an early version of the Review of Particle Physics.

status of the Eightfold Way at the time of the discoveries. In addition, when the η was first reported there were other theoretical explanations for that particle.[30] Although the Eightfold Way was proposed in early 1961 it does not seem to have been known to Pevsner's group when they reported the η in late 1961. When Bastien et al. identified the η as the remaining vacancy in the pseudoscalar meson octet in early 1962 they remarked that they had had discussions with, and assistance from, Gell-Mann. It was not until the announcement of the discovery of the Ξ(1530) particles at the CERN conference in July 1962, along with discussion by Gell-Mann that the Eightfold Way became a viable candidate for a theory of elementary particles. By that time, the η was regarded as "old evidence."

Several philosophers of science have argued that "old evidence," evidence known at the time a theory of hypothesis was proposed, cannot provide support for that hypothesis. The fear is that the theory has been adjusted to fit that evidence. This view seems incorrect, and is also at odds with the history of physics. Many theories or hypotheses are proposed to explain existing data or phenomena. Max Planck, for instance, introduced the quantum to explain the spectrum of blackbody radiation. Kepler's Laws of Planetary Motion were known well before they were explained in Isaac Newton's *Principia*, and certainly provided support for Newton's Laws of Motion and his Law of Universal Gravitation.[31] This is also closely related to arguments concerning the support provided by either prediction and accommodation. Once again, some philosophers of science have argued that evidence that is predicted provides more support than evidence that is accommodated. In this view the Ω^- was a prediction, whereas the η would be regarded as an accommodation.[32] Yuval Ne'eman argues against this.

> The eightfold was celebrated a victory [the discovery of the Ω^-]. With the aid of its diagrams many numerical results were predicted which were later verified experimentally. However, the accomplishment most remembered and acclaimed was the discovery of the omega minus. *It was compared to the three missing elements in Mendeleev's periodic table, whose properties Mendeleev had been able to predict.* Actually, the importance attached to a successful prediction is associated with human psychology rather than with the scientific methodology. It would not have detracted at all from the effectiveness of the eightfold way if the Ω^- had been discovered *before* the theory was proposed. But human nature stands in great awe when a prophecy comes true, and regards the realization of a theoretical prediction as an irrefutable proof of the validity of the theory (Ne'eman and Kirsh 1986, p. 202, first emphasis added, second emphasis in the original.).

Although we believe that "irrefutable proof" is too strong a statement, we do believe that this may be an explanation for the difference in treatment of the discoveries of the η meson and the Ω^- hyperon.[33]

---

[30]See, for example, Nambu (1957) and Sakurai (1960a, 1960b, 1961)

[31]For a discussion of the philosophical issues in this episode see Franklin and Howson (1985) and the references cited.

[32]For a philosophical discussion of this issue and an argument that that prediction does not provide more support for a hypothesis than accommodation see Howson and Franklin (1991).

[33]Another oddity was the1954 report of the "Possible Existence of a New Hyperon" by Eisenberg (1954) Interestingly Eisenberg named the particle the Ω^-. As Samios later stated, "The Ω^- mass

References

Abrams, G.S., R.A. Burnstein, et al. 1964. Example of decay Ω- → $\Xi^- + \pi^o$. *Physical Review Letters* 13: 670–672.
Aubert, B., R. Barate, et al. 2006. Measurement of the spin of the omega minus hyperon. *Physical Review Letters* 97: 112001–112007.
Barnes, V.E., P.L. Connolly, et al. 1964a. Observation of a hyperon with strangeness minus three. *Physical Review Letters* 12: 204–206.
Barnes, V.E., P.L. Connolly, et al. 1964b. Confirmation of the existence of the omega minus hyperon. *Physics Letters* 12: 134–136.
Bastien, P.L., J.P. Berge, et al. 1962. Decay modes and width of the η meson. *Physical Review Letters* 8: 114–117.
Baubillier, M., I.J. Bloodworth, et al. 1978. A study of the lifetime and spin of omega$^-$ produced in K^- p Interactions at 8.25 GeV/c. *Physics Letters* 78B: 342–346.
Bertanze, L., V. Brisson, et al. 1962. Possible resonances in the $\Xi\pi$ and KK^{bar} systems. *Physical Review Letters* 9: 180–183.
Chew, G.F., M. Gelll-Mann, et al. 1964. Strongly interacting particles. *Scientific American* 210: 74–93.
Deutschmann, M., B. Otter, et al. 1978. Spin and lifetime of the omega$^-$ hyperon. *Physics Letters B* 73B: 96–98.
Dothan, Y., H. Goldberg, et al. 1962. Protonium two-meson annihilation. *Physics Letters* 1: 308–310.
Eden, R.J., and J.R. Taylor. 1963. Resonance multiplets and broken symmetry. *Physical Review Letters* 11: 516–518.
Eisenberg, Y. 1954. Possible existence of a new hyperon. *Physical Review* 96: 541–542.
Fowler, W.B., and N. Samios. 1964. The omega-minus experiment. *Scientific American* 211: 36–45.
Franklin, A., and C. Howson. 1985. Newton and Kepler, A Bayesian approach. *Studies in History and Philosophy of Science* 16: 379–385.
Franklin, A., and R. Laymon. 2019. *Measuring nothing, repeatedly*. San Rafael, CA: Morgan and Claypool.
Frautschi, S. 1963. Broke symmetry abd resonance poles. *Physics Letters* 8: 141–143.
Gell-Mann, M. 1961. *The eightfold way: A theory of strong interaction symmetry*. Pasadena: California Institute of Technology.
Gell-Mann, M. 1962a. Symmetries of baryons and mesons. *Physical Review* 125: 1067–1084.
Gell-Mann, M. 1962b. Strong Interactions of Strange Particles. In *1962 Annual International Conference on High-Energy Physics*, 805.
Gell-Mann, M. 1964. A schematic model of baryons and mesons. *Physics* 8: 214–215.
Gell-Mann, M., and Y. Ne'eman. 2018. *The eightfold way*. Boca Raton: CRC Press.
Gell-Mann, M., and A.H. Rosenfeld. 1957. Hyperons and heavy mesons. *Annual Reviews of Nuclear Science* 7: 407–478.
Goldberg, H., and Y. Ne'eman. 1962. Baryon charge and R-inversion. *Nuovo Cimento* 27: 1–5.
Hemingway, R.J., R. Armenteros, et al. 1978. Omega$^-$ produced in K^-p reactions at 4.2 GeV/c. *Nuclear Physics B* 142: 205–219.
Howson, C., and A. Franklin. 1991. Maher, Mendeleev, and bayesianism. *Philosophy of Science* 58: 574–585.
Lipkin, H.J. 1973. Quarks for pedestrians. *Physics Reports (Section C of Physics Letters)* 8: 173–268.
Nambu, Y. 1957. Possible existence of a heavy neutral meson. *Physical Review* 106: 1366–1367.
Ne'eman, Y. 1961. Derivation of Strong Interactions from a Guage Invariance.
Ne'eman, Y., and Y. Kirsh. 1986. *The particle hunters*. Cambridge: Cambridge University Press.

---

as tabulated in 1992 is 1672 ± 0.3 MeV.... With the precise mass value it is now evident that the Eisenberg event is not an Ω^-. This is because either interpretation of the event $X^- \rightarrow \Lambda K^- +$ 4 MeV or $X^- \rightarrow K^-\Sigma^o$ yield mass values of 1613 MeV and 1690 MeV, respectively, inconsistent with the present mass value (Samios 1993, p. 12)."

Oakes, R.J., and C.N. Yang. 1963. Meson-Baryon resonances and the mass formula. *Physical Review Letters* 11: 174–178.

Pevsner, A., R. Kraemer, et al. 1961. Evidence for a three-pion resonance near 550 Mev. *Physical Review Letters* 7: 421–423.

Pjerrou, G.M. 1962. A resonance in the $\Xi\pi$ system and 1.53 GeV. In *1962 Annual International Conference on High-Energy Physics*, 289–290.

Rosenfeld, A.H. 1975. The particle data group: Growth and operations-eighteen years of particle physics. *Annual Reviews of Nuclear Science* 25: 555–598.

Rosenfeld, A.H., A. Barbaro-Galtieri, et al. 1964. Data on elementary particles and resonant states. *Reviews of Modern Physics* 36: 977–1004.

Sakurai, J.J. 1960a. Theory of strong interactions. *Annals of Physics* 11: 1–48.

Sakurai, J.J. 1960b. Pion resonances. *Il Nuovo Cimento* 16: 388–391.

Sakurai, J.J. 1961. Existence of two $T = 0$ vector mesons. *Physical Review Letters* 7: 355–358.

Samios, N. 1993. Early Baryon and Meson Spectroscopy Culminating in the Discovery of the Omega Minus, Brookhaven National Laboratory.

Samios, N. 1994. Baryon spectroscopy and the omega minus. In: *International Conference on History of Basic Discoveries and Original Ideas in Particle Physics*. Erice, Sicily, BNL-61009.

Samios, N. 1997. Early meson and baryon spectroscopy culminating in the discovery of the omega-minus and charmed baryons. In *The rise of the standard model: Particle physics in the 1960s and 1970s*, ed. L. Hoddeson, L. Brown, M. Riordan, and M. Dresden, 525–541. Cambridge: Cambridge University Press.

Strangeness Minus Three. BBC. https://www.bbc.co.uk/programmes/p01z4p1j.

Zweig, G. 1964a. An SU(3) Model for Strong Interaction Symmetry and Its Breaking I. Geneva, CERN. 8182-Th-401.

Zweig, G. 1964b. An SU(3) model for strong interaction symmetry and its breaking II. In *Developments on the quark theory of hadrons, a reprint collection. Vol. I: 1964-1978*, ed. D. Lichtenberg and S. Rosen, 22–101. Nonantin, MA: Hadronic Press.

Zweig, G. (1980). Origins of the quark model. In *Fourth International Conference on Baryon Resonances*. N. Isgur. Toronto, 439–479.

Chapter 5
Once Should Have Been Enough: Gregor Mendel, "Experiments in Plant Hybridization"

Gregor Mendel is justly regarded as the father of modern genetics. In this chapter we explain why this is so, and why in particular "once was enough" to have established Mendel's pea experiments as a basis and model for the development of genetics. And this despite the fact that R.A. Fisher demonstrated in 1936 by means of a χ^2 analysis that Mendel's results were "too good to be true," that they agreed too well with his hypotheses.

Our presentation proceeds in two steps. First—and putting Fisher's analysis temporarily aside—we review and demonstrate the outstanding quality of Mendel's experimental program. Where that quality was such that Fischer concluded that "[t]he facts available in 1900 were at least sufficient to establish Mendel's contribution as one of the greatest experimental advances in the history of biology" (Fisher 1965).

Second, we review Fisher's χ^2 analysis and examine in some detail the statistics underlying the nature of the "too good to be true" result; what we consider to be the leading responses to that analysis, and whether Mendel's sampling procedures were corrupted by unrecognized confounding causes which led to the data seeming to be too good. We conclude with a consideration of the sense in which it may be said that "once was enough" when it came to Mendel's "Experiments in Plant Hybridization."

5.1 Mendel's Methodology and Experimental Results

The basis for holding Mendel to have been the father of modern genetics is compelling and straightforward. His experiments on pea plants not only suggested the laws of segregation and independent assortment, which are the basis of modern genetics, but also provided strong evidence for those laws. The former law states that variation for contrasting traits is associated with a pair of factors which segregate to individual reproductive cells. The latter states that two or more of these factor-pairs assort independently to individual reproductive cells. The experiments also provided evidence that those factors, which we now call genes, come in two types, dominant

A. Franklin and R. Laymon, *Once Can Be Enough*,
https://doi.org/10.1007/978-3-030-62565-8_5

and recessive. Mendel also provided answers to objections that could have been raised concerning the evidence.

Mendel began his experiments on garden peas (*Pisum sativum* L.) in 1856 and continued them until 1863, a period of approximately eight years. His stated purpose was to investigate whether there was a general law for the formation and development of hybrids, something he noted had not yet been formulated. He stated,

> Those who survey the work in this department will arrive at the conviction that among all the numerous experiments made, not one has been carried out to such an extent and in such a way as to make it possible to determine the number of different forms under which the offspring of hybrids appear, or to arrange these forms with certainty according to their separate generations, or definitely to ascertain their statistical relations.
>
> It requires indeed some courage to undertake a labour of such far-reaching extent; this appears however, to be the only right way by which we can finally reach the solution of a question the importance of which cannot be overestimated in connection with the history of the evolution of organic forms (Bateson 1909, p. 318).[1]

Mendel proposed to remedy the situation and did so. His paper is a model of both clarity and organization. In order to carry out such experiments successfully Mendel required that "The experimental plants must necessarily: (1) Possess constant differentiating characters. (2) The hybrids of such plants must, during the flowering period, be protected from the influence of all foreign pollen or be easily capable of such protection. The hybrids and their offspring should suffer no marked disturbance in their fertility in successive generations (Bateson, p. 319)." He further noted that "In order to discover the relations in which the hybrid forms stand towards each other and towards their progenitors it appears to be necessary that all members of the series developed in each successive generation should be, without exception, subjected to observation (p. 319)."

To persuade the reader of the quality and importance of Mendel's work we will present almost all of Mendel's experiments, his data, and his conclusions. We will also follow the organization of Mendel's paper to illustrate the clarity of his arguments.

A. (F_2) The First Generation [Bred] from the Hybrids.

Mendel began with 34 varieties of peas from which he selected 22 varieties for further experiments. He had confirmed previously, in two years of experimentation, that these varieties bred true. He thus established that his experiments satisfied an important criterion for investigation of the subject. He reported experiments on seven characters that had two easily distinguishable characteristics. These were: (Dominant form listed first)[2]

1. Seed shape: round or wrinkled.

---

[1]All page references to Mendel's paper will refer to the translation contained in Bateson (1909). Other translations are available in Stern and Sherwood (1966, pp. 1–48) and https://plantae.org/experiments-on-plant-hybrids-by-gregor-mendel-new-english-translation/.

[2]Not all of the experiments described in the same section of Mendel's paper necessarily occurred at the same time. Mendel also showed that these characters were dominant and recessive. See discussion below.

2. Cotyledon color: yellow or green.
3. Seed coat color: colored (grey, grey-brown, or leather-brown) or white. The former was always associated with violet flower color and reddish markings at the leaf axils. The latter was associated with white flowers.
4. Pod shape: inflated or constricted.
5. Pod color: green or yellow.
6. Flower position: Axial (along the stem) or terminal (at the end of the stem).
7. Stem length: long (6–7 feet) or short (3/4–1 foot).

The first two are seed characters because they are observed in seed cotyledons, which consist of embryonic tissue. Each seed is thus a genetically different individual and such characters may differ among the seeds produced on a heterozygous plant. Both yellow and green seeds may be observed on a single such plant. One may, in fact, observe these characters for the next generation, without the necessity of planting the seeds. The latter five are plant characters. As William Bateson remarked, "It will be observed that the [last] five are *plant-characters*. In order to see the result of crossing, the seeds must be sown and allowed to grow into plants. The [first] two characters belong to the *seeds* themselves. The seeds of course are members of a generation later than that of the plant which bears them (Bateson 1909, p. 12)." Because of this Mendel would have had a reasonable expectation of what the results of his plant character experiments would be from his observations of the seed characters, before the plants of the next generation were grown.

Mendel's first experiment was to breed a generation of hybrids from his true breeding plants for each of the seven characters. His results for this generation (F_1) clearly showed dominance. He remarked that "In the case of each of the seven crosses the hybrid-character resembles that of one of the parental forms so closely that the other either escapes observation completely or cannot be detected with certainty (Bateson, p. 324)."

He then allowed these monohybrids to self-fertilize. He found a 3:1 ratio for plants that showed the dominant character to those that possessed the recessive character in this F_2 generation. He found that this ratio held for all the characters observed in the experiments and that "*transitional forms were not observed in any experiment* (p. 326)." His results are shown in Table 5.1. He concluded, "If now the results of the whole of the experiments be brought together, there is found, as between the number of forms with the dominant character and recessive characters, an average ratio of 2.98–1, or 3–1 (p. 328)." In modern notation, self-fertilization of a heterozygous plant is genetically equivalent to the cross Aa x $Aa \rightarrow AA + 2Aa + aa$. Both the AA and Aa plants will display the dominant character, whereas an aa plant will display the recessive character,[3] thus giving a 3:1 ratio for dominant to recessive.

Mendel noted that the distribution of characters varied in both individual plants and in individual pods. He illustrated this with data from the first ten plants in the seed character experiments (Table 5.2). The variation in both the ratios of the characters and in the number of seeds per plant is considerable. Mendel also presented the

[3]Mendel used A, Aa, and a, respectively to denote genotypes we currently symbolize as *AA*, *Aa*, and *aa*.

Table 5.1 Mendel's results for the F_2 generation of monohybrid experiments

Trait	Dominant	Number	Recessive	Number	Ratio
1. Seed shape	Round	5474	Angular	1859	2.96
2. Cotyledon color	Yellow	6022	Green	2001	3.01
3. Seed coat color	Colored	705	White	224	3.15
4. Pod shape	Inflated	882	Constricted	299	2.95
5. Pod color	Green	428	Yellow	152	2.82
6. Flower position	Axial	651	Terminal	207	3.14
7. Stem length	Long	787	Short	277	2.89
Total	Dominant	14,949	Recessive	5010	2.98

From Mendel 1865 in Bateson (1909, p. 328)

Table 5.2 Mendel's Results for the first 10 plants in the experiments on seed shape and seed color

Plant	Experiment 1		Experiment 2	
	Shape of seeds		Coloration of albumen	
	Round	Wrinkled	Yellow	Green
1	45	12	25	11
2	27	8	32	7
3	24	7	14	5
4	19	10	70	27
5	32	11	24	13
6	26	6	20	6
7	88	24	32	13
8	22	10	44	9
9	28	6	50	14
10	25	7	44	18
Ratio	3.33:1		3.08:1	

From Mendel 1865 in Bateson (1909, p. 327)
Note The fact that the number of seeds in each plant differs for each numbered plant shows clearly that the plants for Experiments 1 and 2 are different plants. Thus, plant 1 in Experiment 1 has 57 seeds, whereas plant 1 in Experiment 2 has 36 seeds

extreme variations. "As extremes in the distribution of the two seed characters in one plant, there were observed in Expt. 1, an instance of 43 round and only 2 wrinkled, and another of 14 round and 15 wrinkled seeds. In Expt. 2 there was a case of 32 yellow and only 1 green seed, but also one of 20 yellow and 19 green (p. 327)." Mendel was clearly willing to present data that deviated considerably from his expectations. Mendel noted that care must be taken in these experiments. He stated that, "These two experiments are important for the determination of the average ratios, because with a smaller number of experimental plants they show that very considerable fluctuations may occur. In counting the seeds, also, especially in Expt. 2, some care is requisite, since in some of the seeds of many plants the green colour of the albumen is less developed, and at first may be easily overlooked.... In luxuriant plants this appearance was frequently noted. *Seeds which are damaged by insects during their development often vary in colour and form, but, with a little practice in sorting, errors are easily avoided* (pp. 327–328, emphasis added)." Mendel was remarking on the statistical significance of his results. He also noted a possible problem with his classification of seeds and stated how it could be solved.

B. (F_3) The Second Generation [Bred] from the Hybrids.[4]

At the end of the section describing the first-generation experiments Mendel remarked that the observed dominant character could have a "*double significance.*" It could be either a pure parental (dominant) character or a hybrid character. "In which of the two significations it appears in each separate case can only be determined by the following generation. As a parental character it must pass over unchanged to the whole of the offspring; as a hybrid-character, on the other hand, it must maintain the same behavior as in the first generation (p. 329)."[5] He further noted that those plants that show the recessive character in the first generation (F_2) do not vary in the second generation (F_3). They breed true. That was not the case for those plants showing the dominant character. "Of these, *two*-thirds yield offspring which display the dominant and the recessive characters in the proportion of 3–1, and thereby show exactly the same ratio as the hybrid forms, while only *one*-third remains with the dominant character (p. 329)." In other words, of those F_2 generation plants showing the dominant character two thirds were heterozygous (Aa), or hybrid, and one third homozygous (AA). For the seed characters Mendel reported the following results: (1) From 565 plants raised from round seeds 372 produced both round and wrinkled seeds in the proportion of 3–1 whereas 193 yielded only round seeds, a ratio of 1.93–1. For plants raised from yellow seeds 353 yielded both yellow and green seeds in the proportion 3–1 whereas 166 yielded only yellow seeds, a ratio of 2.13–1.

The experiments on plant characters required more effort. "For each separate trial in the following experiments [on plant characters] 100 plants were selected which displayed the dominant character in the first generation [F_2], and in order to ascertain the significance of this, 10 [F_3] seeds of each plant were cultivated (p. 330)."[6] A plant was classified as homozygous if all of the 10 offspring had the dominant

[4]These are the first experiments on the 2–1 ratio that would be of concern to Fisher (see Footnote 6). A second instance is the trifactorial hybrid experiment, discussed below.

[5]This is true for the plant characteristics. The seed characteristics would appear in the same generation.

[6]There is considerable discussion in the ensuing controversy over whether the German word should be translated as "sown" or "cultivated." In their translation Stern and Sherwood (1966) used "sown." The choice of 10 seeds is also of significance. Fisher would later argue that because only 10 seeds were planted that there is a 5.6% probability that heterozygous plants would be classified as homozygous and thus be undercounted. Because of this he argued that the ratio should be 1.8874–1.1126, or about 1.7–1, rather than 2–1.

Table 5.3 Mendel's Results for the heterozygous-homozygous experiment (the 2–1 experiment)

Experiment	Dominant	Hybrid
3. Seed coat color (grey-brown or white)	36	64
4. Pod shape (smooth or constricted)	29	71
5. Pod color (green or yellow)	40	60
6. Flower location (axillary or terminal)	33	67
7. Stem length (long or short)	28	72
8. Repetition of Experiment 5	35	65

From Mendel 1865 in Bateson (1909, p. 330)

character and classified as heterozygous otherwise.[7] Mendel's results for the plant characteristics are shown in Table 5.3. Mendel noted that the first two experiments on seed characters were of special importance because of the large number of plants that could be compared. Those experiments yielded a total of 725 hybrid plants and 359 dominant plants which, "gave together almost exactly the average ratio of 2–1 (p. 330)." Experiment 6 also yielded almost the exact ratio expected, whereas for the other experiments, as Mendel noted, "the ratio varies more or less, as was only to be expected in view of the smaller number of 100 trial plants (p. 330)." Mendel was, however concerned about Experiment 5 (the color of unripe pods), whose result was 60–40, which he regarded as deviating too much from the expected 2–1 ratio.[8] He repeated the experiment and obtained a ratio of 65–35, and was satisfied that, "*The average ratio of 2–1 appears, therefore, as fixed with certainty* (p. 330)." Mendel did not attempt to hide any of his results and, in particular, those that deviated from his expectations, because he presented the results for both the original Experiment 5 as well as its replication. The sum totals for the six plant characteristic experiments, including the repetition of Experiment 5, was 399 (hybrid) to 201 (dominant), or 1.99–1.

Mendel's conclusion was quite clear:

> The ratio of 3–1 in accordance with which the distribution of the dominant and recessive characters results in the first generation, resolves itself into a ratio of 2:1:1 if the dominant character be differentiated according to its significance as a hybrid-character or as a parental one. Since the members of the first generation [F_2] spring directly from the seed of the hybrids [F_1], *it is now clear that the hybrids form seeds having one or the other of the two differentiating characters, and of these one half develop again the hybrid form, while the other half yield plants which remain constant and receive the dominant or the recessive characters, [respectively], in equal numbers* (pp. 330–331).

C. The Subsequent Generations [Bred] from the Hybrids.

Mendel suspected that the results he had obtained from the first and second generations produced from monohybrids were probably valid for all of the subsequent progeny. He continued the experiments on the two seed characters, shape and color,

[7] This is a reasonable reading of what Mendel wrote, and it was R.A. Fisher's interpretation. There is considerable discussion in the later literature about whether Fisher was correct.

[8] A modern statistician would not regard this as a significant deviation.

for six generations, the experiments on seed coat color and stem length for five generations, and the remaining three experiments on pod shape, color of pods, and position of flowers, for four generations "... and no departure from the rule has been perceptible. The offspring of the hybrids separated in each generation in the ratio of 2:1:1 into hybrid and constant forms [pure dominant and pure recessive] (p. 331)." He did not, however, present his data for the experiments on the subsequent generations.[9] He went on to state, "If A be taken as denoting one of the two constant characters, for instance, the dominant, a, the recessive, and Aa the hybrid form in which both are conjoined, the expression A + 2Aa + a shows the terms in the series for the progeny of the hybrids of two differentiating characters (p. 331)."

D. The Offspring of Hybrids in Which Several Differentiating Characters are Associated.

Mendel's next task, as he put it, was to investigate whether the laws he had found for monohybrid plants also "applied to each pair of differentiating characters when several diverse characters are united in the hybrid by crossing (p. 332)."

He went on to describe the experiments. "Two experiments were made with a considerable number of plants. In the first experiment the parental plants differed in the form of the seed and in the colour of the albumen; in the second in the form of the seed, in the colour of the albumen, and in the colour of the seed-coats. Experiments with seed characters give the result in the simplest and most certain way (p. 333)." He was no doubt referring to the greater number of seeds than plants, which provides data with greater statistical significance, and also to the fact that the shape of the seeds and the color of albumen (cotyledons) could be seen in the second generation, without the need to plant a third generation.

In these experiments Mendel distinguished between the differing characters in the seed plant and the pollen plant. A, B, and C represented the dominant characters of the seed plant and a, b, c the recessive characters of the pollen plant, with hybrids represented as Aa, Bb, and Cc.[10]

1. First Experiment (Bifactorial)

Mendel's first experiment used two seed characters in which the seed plant (AB) was A (round shape) and B (yellow cotyledon), and the pollen plant (ab) was a (wrinkled shape) and b (green albumen). The fertilized seeds were all round and yellow, as expected. He then raised plants from these seeds and obtained 15 plants with 556 seeds distributed as follows:

315 round and yellow
101 wrinkled and yellow
108 round and green

[9]Mendel did not present all of his data from several of his other experiments.

[10]In other experiments, discussed below, Mendel would investigate whether there was any difference in results depending on which characters were associated with the seed and pollen plants, respectively. He would conclude that there was no difference.

Table 5.4 Mendel's results for the bifactorial experiment

	A (round)	Aa (hybrid)	a (angular)
B	AB (round, yellow) 38	AaB (round yellow and angular yellow) 60	aB (angular, yellow) 28
Bb	ABb (round yellow and green) 65	AaBb (round yellow and green and angular yellow and green) 138	aBb (angular yellow and green 68
b	Ab (round green) 35	Aab (round and angular green) 67	ab (angular green) 30

From Mendel 1865 in Bateson (1909, pp. 333–34)

32 wrinkled and green[11]

All of these seeds were planted in the following year and Mendel's results are shown in Table 5.4. Mendel separately recorded the results for each set of the 556 seeds (i.e. round and yellow, round and green, wrinkled and yellow, wrinkled and green). He noted that there were nine different forms (we would say genotypes) and a summary of his results appears in Table 5.4.

> The whole of the forms may be classed into three essentially different groups. The first includes those with the signs AB, Ab, aB, ab: they possess only constant characters and do not vary again in the next generation. Each of these forms is represented on the average thirty-three times. The second group includes the signs ABb, aBb, AaB, Aab: these are constant in one character and hybrid in another, and vary in the next generation only as regards the hybrid character. Each of these appears on average sixty-five times. The form AaBb occurs 138 times: it is hybrid in both characters, and behaves exactly as do the hybrids from which it is derived
>
> If the numbers in which the forms belonging to these classes appear to be compared, the ratios of 1, 2, 4 are unmistakably evident. The numbers 33, 65, 138 present very fair approximations to the numerical proportions 33, 66, 132 (p. 334)."

Mendel had a very good feel for his data and an ability to see the underlying patterns in his results despite statistical fluctuations. Mendel concluded that these results "indisputably" showed that the results could be explained by the combination of A + 2Aa + a and B + 2Bb + b (i.e., AB + 2AaB + aB + 2ABb +4AaBb + 2aBb + Ab + 2Aab + ab). Of these the AB, the two AaB, the two ABb, and the four AaBb, a total of nine show both dominant traits. The aB and two aBb show one dominant trait, as does the Ab and the two Aab. Both a show a total of three for the dominant B and A traits, respectively, and ab gives a total of one, hence the 9:3:3:1 ratios.

2. Second Experiment (Trifactorial)

[11]These results are a good fit to the expected 9:3:3:1 ratio. See discussion below.

Table 5.5 Mendel's results for the trifactorial experiment

8 plants ABC	22 plants ABCc	45 plants ABbCc
14 plants ABc	17 plants AbCc	36 plants aBbCc
9 plants AbC	25 plants aBCc	38 plants AaBCc
11 plants Abc	20 plants abCc	40 plants AabCc
8 plants aBC	15 plants ABbC	49 plants AaBbC
10 plants aBc	18 plants ABbc	48 plants AaBbc
10 plants abC	19 plants aBbC	
7 plants abc	24 plants aBbc	
	14 plants AaBC	78 plants AaBbCc
	18 plants AaBc	
	20 plants AabC	
	16 plants Aabc	

From Mendel 1865 in Bateson (1909, p. 335)

In this experiment Mendel investigated whether the results he had obtained in both the monohybrid and bifactorial experiments held for an experiment in which three different characters were examined, the trifactorial experiment. He remarked, "Among all the experiments it demanded the most time and trouble (p. 335)." The characters investigated were: Seed plant (ABC); A round shape, B yellow albumen, and C grey-brown seed coat; pollen plant (abc); a wrinkled seed, b green albumen, and c white seed coat. The first two were seed characters and could be observed immediately, whereas seed coat color, a plant character, required plants from the next generation. Mendel obtained 687 seeds from 24 hybrid plants, from which he successfully grew 639 plants and "as further investigations showed," he obtained the results depicted in Table 5.5. He summarized his data as follows

> The whole expression contains 27 terms. Of these 8 are constant in all characters, and each appears on the average 10 times; 12 are constant in two characters and hybrid in the third; each appears on the average 19 times; 6 are constant in one character, hybrid in the other two; each appears on the average 43 times. One form appears 78 times and is hybrid in all of the characters. The ratios 10, 19, 43, 78 agree so closely with the ratios 10, 20, 40, 80, or 1, 2, 4, 8, that this last undoubtedly represents the correct value. (p. 336)[12]

Mendel also stated that the series resulted from combining, A + 2Aa + a, B + 2Bb + b, and C + 2Cc + c. He had strong feeling about the results expected and was willing to accept conclusions despite limited statistics. As Fisher remarked, "He evidently felt no anxiety lest his counts should be regarded as insufficient to prove his theory (Fisher 1936, p. 119)."

Mendel remarked that he had conducted several other experiments in which the remaining characters were combined in twos and threes and which gave approximately equal results, but he presented none of his data for these experiments.

[12]Notice again the excellent feel that Mendel has for his data. He sees the significant pattern and neglects the small deviations from that pattern. This is true of all of his analyses.

> There is therefore no doubt that for the whole of the characters involved in the experiments the principle applies that *the offspring of the hybrids in which several essentially different characters are combined exhibit the terms of a series of combinations, in which the developmental series for each pair of differentiating characters are united.* It is demonstrated at the same time that *the relation of each pair of different characters in hybrid union is independent of the other differences in the two original parental stocks* (pp. 336–337)

Mendel also thought that his results justified belief that the same behavior applied to characters which could not be so easily distinguished. He noted, however, the difficulty of such experiments. "An experiment with peduncles of different lengths gave on the whole a fairly satisfactory result, although the differentiation and serial arrangement of the forms could not be effected with that certainty which is indispensable for correct experiment (p. 338)."

E. The Reproductive Cells of Hybrids

In his bifactorial and trifactorial experiments Mendel used seed plants with the dominant characters and pollen plants with the recessive character. The question remained whether his results would remain the same if those parental types were reversed. He also stated that in hybrid plants it was reasonable to assume that there were as many kinds of egg and pollen cells as there were possibilities for constant combination forms. He further noted that this assumption combined with the idea that the different kinds of egg and pollen cells are produced on average in equal numbers would explain all of his previous results.

Mendel proposed to investigate these issues explicitly in a series of experiments. He chose true breeding plants as follows: Seed plant AB, where A and B were round shape and yellow albumen, respectively; pollen plant ab, where a and b were wrinkled shape and green albumen, respectively. These were artificially fertilized and the hybrid AaBb obtained. Both the artificially fertilized seeds, together with several seeds from the two parental plants, were sown. He then performed the following fertilizations:

1. The hybrids with the pollen from AB.
2. The hybrids with the pollen from ab.
3. AB with pollen of the hybrid.
4. ab with pollen of the hybrid.

For each of these experiments all of the flowers on three plants were fertilized. Mendel stated that if his assumptions were correct then the hybrids would contain egg and pollen cells of the form AB, Ab, aB, and ab. When combined with the egg and pollen cells from the parental plants AB and ab the following patterns emerge.

1. AB, ABb, AaB, AaBb.
2. AaBb, Aab, aBb, ab.
3. AB, ABb, AaB, AaBb.
4. AaBb, Aab, aBb, ab.

These genotypes should occur with equal frequency in each experiment. Experiments 1 and 3, as well as experiments 2 and 4, would demonstrate that the results are independent of which parent is used for pollen and which is used for seed. Mendel also noted that there would be statistical fluctuations in his data.

> If, furthermore, the several forms of the egg and pollen cells of the hybrids were produced on an average in equal numbers, then in each experiment the said four combinations should stand in the same ratio to each other. A perfect agreement in the numerical relations was, however, not to be expected, since in each fertilization, even in normal cases some egg cells remain undeveloped or subsequently die, and even many of the well-formed seeds fail to germinate when sown. The above assumption is also limited in so far that, while it demands the formation of an equal number of the various sorts of egg and pollen cells, it does not require that this should apply to each separate hybrid with mathematical exactness (p. 340).

Mendel predicted that in experiments 1 and 3 all of the seeds produced would be round and yellow, the result of dominance. For experiments 2 and 4 his expectations were that round yellow seeds, round green seeds, wrinkled yellow seeds, and wrinkled green seeds would be produced in equal proportions. "The crop fulfilled these expectations perfectly (p. 341)." Experiments 1 and 3 produced 98 and 94 round and yellow seeds exclusively, respectively. Experiment 2 produced 31 round yellow seeds, 26 round green seeds, 27 wrinkled yellow seeds, and 26 wrinkled green seeds. Experiment 4 produced 24 round yellow seeds, 25 round green seeds, 22 wrinkled yellow seeds, and 27 wrinkled green seeds. "There could scarcely be now any doubt of the success of the experiment; the next generation must afford the final proof (p. 341)."

Mendel sowed all of the seeds obtained in the first experiment and 90 plants from 98 seeds bore fruit and in the third experiment 87 plants from 94 seeds bore fruit. "In the second and fourth experiments the round and yellow seeds yielded plants with round and wrinkled yellow and green seeds, AaBb. From the round green seeds plants resulted with round and wrinkled green seeds, Aab. The wrinkled yellow seeds gave plants with wrinkled yellow and green seeds, aBb. From the wrinkled green seeds plants were raised which yielded again only wrinkled green seeds, ab. (pp. 341–342)." Mendel's results are also shown in Table 5.6. He concluded that, "In all the experiments, therefore, there appeared all the forms which the proposed theory demands, and they came in nearly equal numbers (p. 342)."

Mendel conducted a second set of experiments to test his assumptions. In these experiments he made selections so that each character should occur in half the plants if his assumptions were correct. In these experiments A conferred violet-red flowers, a conferred white flowers, B long stems, and b short stems. He fertilized Ab (violet-red flowers, short stem) with ab (white flowers, short stem) producing hybrid Aab. In addition, aB (white flowers, long stem) was also fertilized with ab, yielding hybrid aBb. In the second year the hybrid Aab was used as the seed plant and hybrid aBb as pollen plant This should produce the combinations AaBb, aBb, Aab, and ab. In the third year half the plants would have Aa (violet-red flowers), half a (white flowers), half Bb (long stems), and half b (short stems). The results are shown in Table 5.7. Mendel modestly concluded, "The theory adduced is therefore satisfactorily confirmed in this experiment also (p. 343)."

Table 5.6 Mendel's results from the gametic experiments 1 and 3 and experiments 2 and 4

Experiments			
1	3		
20	25	Round yellow seeds	AB
23	19	Round yellow and green seeds	ABb
25	22	Round and angular yellow seeds	AaB
22	21	Round and angular, yellow and green seeds	AaBb

Experiments			
2	4		
31	24	Yielded plant seeds of form	AaBb
26	25	Yielded plant seeds of form	AaB
27	22	Yielded plant seeds of form	aBb
26	27	Yielded plant seeds of form	ab

From Mendel 1865 in Bateson (1909, pp. 341–42)

Table 5.7 Mendel's results for the flower color-stem length experiments

Term	Color of flower	Stem	Number
AaBb	Purplish-red	Long	47
aBb	White	Long	40
Aab	Purplish-red	Short	38
Ab	White	Short	41

Trait	Number
Purplish-red flower color (Aa)	85 plants
White flower color (a)	81 plants
Long stem (Bb)	87 plants
Short stem (b)	79 plants

From Mendel 1865 in Bateson (1909, p. 343)

Mendel also performed other experiments, with fewer plants, on pod shape, pod color, and flower position, and "the results were in full agreement." No numerical data were presented.

Mendel concluded, "Experimentally, therefore, the theory is confirmed that the *pea hybrids form egg and pollen cells which, in their constitution, represent in equal numbers all constant forms which result from the combination of characters united in fertilization* (p. 343)." He also stated, "It was furthermore shown by the whole of the experiments that it is perfectly immaterial whether the dominant character belong to the seed-bearer or to the pollen-parent; the form of the hybrid remains identical in both cases (p. 325)."[13]

[13]Mendel remarked that this phenomenon had also been emphasized by Gärtner.

Almost all of Mendel's published numerical data from his pea experiments have been presented. In discussing these results Mendel again demonstrated that he understood, at least qualitatively, the statistical nature of his data. He stated

> This represents the average results of the self-fertilisation of the hybrids when two differentiating characters are united in them. In individual flowers and in individual plants, however, the ratios in which the forms of the series are produced may suffer not inconsiderable fluctuations. Apart from the fact that the numbers in which both sorts of egg cells occur in the seed vessels can only be regarded as equal on the average, it remains purely a matter of chance which of the two sorts of pollen may fertilise each separate egg cell. For this reason the separate values must necessarily be subject to fluctuations, and there are even extreme cases possible, as were described earlier in connection with the experiments on the form of the seed and the colour of the albumen. The true ratios of the numbers can only be ascertained by an average deduced from the sum of as many single values as possible; the greater the number the more are merely chance effects eliminated (p. 345).[14]

Despite the quality of Mendel's work, its significance was not understood until 1900 when his work was replicated by de Vries (1900), Correns (1900), and by von Tschermak (1900). His work was not, however, completely ignored. R.A. Fisher noted that the journal in which Mendel published his results was reasonably well known and was widely distributed. It was received by both the Royal Society and the Linnaean Society.

Mendel also corresponded with Carl von Nägeli, a leading botanist. And sent him a copy of his paper. "The acknowledged pre-eminence your honor enjoys in the detection and classification of wild-growing plant hybrids makes it my agreeable duty to submit for your kind consideration the description of some experiments in artificial fertilization (Letter from Mendel to von Nägeli, December 31, 1866, quoted in Stern and Sherwood (1966, p. 56)." Von Nägeli did not seem to be impressed and in a later letter Mendel defended his results. "I am not surprised to hear from your honor speak of my experiments with mistrustful caution; I would not do otherwise in a similar case. Mendel felt obligated to answer two of von Nägeli's criticisms. The first concerned the constancy of Mendel's results over several generations. Mendel remarked that he had, in fact, extended his measurements over several generations, as described earlier. "The second point, on which I wish to elaborate briefly, contains the following statement: 'You should regard the numerical expressions as being only empirical, because they cannot be proved rational (p. 61).'"

> My experiments with single traits all lead to the same result: that from the seeds of hybrids, plants are obtained half of which in turn carry the hybrid trait (Aa), the other half, however, receive the parental traits A and a in equal amounts. Thus, on the average, among four plants two have the hybrid trait (Aa), one the parental trait A, and the other the parental trait a. Therefore 2Aa + A + a or A + 2Aa + a is the empirical series for the two differing traits. Likewise, it was shown in an empirical manner that, if two or three differing traits are combined in the hybrid, the series is a combination of two or three simple series. Up to this point I don't believe I can be accused of having left the realm of experimentation. If then I extend this combination of simple series to any number of differences between the two

[14]Mendel published a second paper on hybrids in Hieracium (Mendel 1870) His results disagreed with his earlier work. One can explain this because Hieracium reproduces by apomixis, the replacement of the normal sexual reproduction by asexual reproduction, without fertilization.

parental plants, I have indeed entered the rational domain. This seems permissible, however, because I have proved by previous experiments that the development of a pair of differing traits proceeds independently of any other differences. (Letter from Mendel to von Nägeli, April 18, 1867, quoted in Stern and Sherwood (1966, p. 63).

R.A. Fisher later noted that von Nägeli was either unimpressed by Mendel's results or anxious to warn students against paying attention to them. Simon Mawer was less sympathetic to von Nägeli. In his book *Gregor Mendel: planting the seeds of genetics* (2006), He gives us an interesting and detailed account of Nägeli's correspondence with Mendel. Mawer underlines that, at the time Nägeli was writing to Mendel, Nägeli "must have been preparing his great work entitled *A mechanico-physiological theory of organic evolution* (published in 1884, the year of Mendel's death) in which he proposes the concept of the 'idioplasm' as the hypothetical transmitter of inherited characters" Mawer notes that, in this Nägeli book, there is not a single mention of the work of Gregor Mendel. That prompted him to write: "We can forgive von Nägeli for being obtuse and supercilious. We can forgive him for being ignorant, a scientist of his time who did not really have the equipment to understand the significance of what Mendel had done despite the fact that he (von Nägeli) speculated extensively about inheritance. But omitting an account of Mendel's work from his book is, perhaps, unforgivable." (Mawer 2006, p. 81).

Fisher also cited Focke who, in his *Pflanzenmischlinge* (1881), made no fewer than fifteen references to Mendel's work. As Fisher made clear, Focke did not understand Mendel's work and seemed to prefer the more comprehensive contributions of Kolreuter, Gärtner, and others. Fisher remarked that Focke "had overlooked in his chosen field, experimental researches conclusive in their results, faultlessly lucid in presentation, and vital to the understanding not of one problem of current interest, but of many (p. 137)."

5.2 The Rediscovery

In 1900 Mendel's experiments were replicated by Carl Correns, Hugo de Vries, and Erich von Tschermak, with statements crediting both Mendel's priority and the quality of his work. Correns conclusions included

1. *Of the two antagonistic characteristics, the hybrid carries only one,* and that in complete development. Thus, in this respect the hybrid is indistinguishable from one of the two parents. There are no transitional forms.
2. In the formation of pollen and ovules the two antagonistic characteristics separate, following for the most part simple laws of probability.

These two statements, in their most essential points, were drawn up long ago by Mendel for a special case (peas).[footnote by de Vries here] These formulations have been forgotten and their significance misunderstood.[footnote by de Vries here] As my experiments show, they possess generalized validity for true hybrids ((de Vries 1900), quoted in Stern and Sherwood, p. 110).

The first footnote cites Mendel's original paper and notes that, "This important paper is so seldom cited that I first learned of its existence after I had completed the majority of my experiments and had deduced from them the comments communicated in the text." The second footnote remarking on the misunderstanding of Mendel's work cited Focke's *Die Pflanzenmischlinge..*"

Correns was equally explicit in crediting Mendel. The title of his paper was "G. Mendel's Law Concerning the Behavior of Progeny of Varietal Hybrids."

> In my hybridization experiments with varieties of maize and peas, I have come to the same results as de Vries, who experimented with varieties of many different kinds of plants, among them two varieties of maize. When I discovered the regularity of the phenomena, and the explanation thereof – to which I shall return presently – the same thing happened to me now seems to be happening to de Vries: I thought I had found *something new. But then I convinced myself that the Abbot Gregor Mendel in Brunn, had, during the sixties, not only obtained the same result through extensive experiments with peas, which lasted for many years, as did de Vries and I, but had also given exactly the same explanation, as far as that was possible* in 1866.
>
> (Correns 1900) quoted in Stern and Sherwood, pp. 119–120).

Correns was also critical of Focke. "Mendel's paper, although mentioned, is not properly appreciated in Focke's Die Pflanzen-Mischlinge, and which otherwise has hardly been noticed, is among the best that has ever been written about hybrids,... ((Correns (1900) quoted in Stern and Sherwood, p. 120)."

Tschermak's work (1900) was not as insightful as that of Correns, de Vries, or Mendel. As Stern and Sherwood remark, "Tschermak's papers of 1900 not only lack fundamental analysis of his breeding results but clearly show that he had not developed an interpretation (Stern and Sherwood 1966, p. xi)." Tschermak seems to admit this in a letter to H.F. Roberts, published in 1929. He states, "quite intentionally, I expressed the [rules of inheritance] at first purely descriptively or phenomenologically, in order not at once to anchor the newly-beginning experimental phase of the doctrine of inheritance ... to definite theoretical terms (Tschermak letter to H.F. Roberts, quoted in Stern and Sherwood, p. xi)." As discussed earlier, Mendel had done exactly that. As he wrote to von Nageli, he had "indeed entered the rational domain." Stern and Sherwood note, however, that "Tschermak's influence on the recognition of Mendelian genetics by plant breeders was considerable (Stern 1966, p. xi)."

5.3 "Too Good to Be True"

In 1936 R.A. Fisher, the distinguished British geneticist and statistician, reanalyzed Mendel's data and concluded, on the basis of χ^2 analysis, that Mendel's data fit his hypotheses too well (Fisher 1936). They were "too good to be true." Fisher obtained a value of $\chi^2 = 41.6056$ for Mendel's 84 experiments. A fit this good between data and theory has a probability of 0.00007, a highly unlikely event. Given his admiration for Mendel's work, Fischer suggested that Mendel's data had been falsified by a well-intentioned, unnamed assistant. Fisher's χ^2 analysis, however, attracted little attention until around 1965, the centenary of the publication of Mendel's paper. This led to an avalanche of commentary that has been extensively reviewed in Franklin et al. (2008, pp. 1–77) which also contains reprints of several of the most important papers in that deluge.

Here we will focus attention on what we consider in retrospect to be the most salient responses to Fischer's conclusion that Mendel's data were "too good to be true." We do this to indicate in a telling and concise manner the nature of the problems with systematic uncertainty and possible confounding causes that Mendel—in large part unknowingly—had to deal with. Once Mendel's work had been rediscovered these problems would in turn provide an ongoing challenge for the development of the science of genetics.

In short, there are two basic approaches to dealing with what has become known as the Mendelian Paradox, that is, the conflict between Mendel's beautifully conceived experimental scheme and the surprising agreement between his data and his hypotheses. First, that Mendel somehow, perhaps unknowingly and inadvertently, biased his experimental procedures by improperly culling suspect progeny from his test samples. Second, that the expression of the Mendelian factors was corrupted by confounding causes which meant that Mendel's experimental samples were not extracted from a population where those factors satisfied the binomial distribution. In other words, Fisher's χ^2 analysis was inapplicable because Mendel's peas were either not independently distributed within self-fertilizing pods or because of some confounding cause were not randomly expressed.

We begin our review with the second approach because, in the present context, it best sets the stage for a consideration of the first approach. Teddy Seidenfeld considered such a confounding process based on what he refers to as the Correlated Pollen Model[15]:

> Suppose that within the pea-flower for hybrids, 10 egg cells form according an i.i.d. [independent and identically distributed] "fair" (binomial) distribution. However, approximating the speculated, checkerboard pattern that pollen have on the anther, suppose that exactly 5 of every 10 pollen cells arriving at the egg cells are dominant. Last, assume that, with equal probability, 2 of these 10 zygotes spontaneously abort, leaving 8 peas/pod. The result is a model where pollen cells are negatively correlated within a pod (Seidenfeld in Franklin et al. (2008, p. 233).

He then incorporated this—admittedly somewhat speculative model of pollination—into a test hypothesis with the result that a χ^2 analysis yields a variance of 74% of that expected for a binomial distribution. This reduced the corresponding probability to approximately 0.96. Still quite large, but not the extraordinary value of 0.99993 obtained by Fisher. Accordingly, any charge of undue manipulation by Mendel would, at the least, have to be mitigated. Guilty perhaps, but only to a lesser offense.

---

[15]Seidenfeld was unaware that this model had been proposed earlier by Beadle (1967) and by Thoday (1966).

But alas, Seidenfeld's ingenious alternative test hypothesis does not work for Mendel's experiments on gametic ratios in Sect. 9 of his paper. Disappointing but not surprising because the gametic experiments reported "the results of artificial fertilizations, all, where Mendel 'dusted' the pollen onto each stigma for the flowers" (Seidenfeld in Franklin et al. (2008), p. 241).

While not entirely successful, Seidenfeld's Correlated Pollen Model does serve to highlight many of the problems with what is now known as systematic uncertainty that confronted Mendel. In particular, it provides an instance where Mendel's "factor" hypothesis can be separated from the actual test environment in which those factors are (again to make use of modern terminology) expressed in that environment. Here that expression is distorted by the pollen producing mechanism (proposed in the Correlated Pollen Model) such that the resulting distribution of the factors as expressed is not the otherwise expected binomial distribution.

This sort of confounding may be characterized as "phenotypic" insofar as it manifests itself in the external and "directly" observable configuration of the phenotype. Thus, it's not as if the Mendel's sampling procedure was defective or fraudulent but rather that the total space of expressed factors from which the sample had been extracted is not the binomial distribution of what would have been the otherwise non-confounded Mendelian factors. Contrast this sort of phenotypic confounding with the possible sorts of "genotypic" confounding of the expression of the Mendelian factors at the chromosome level. So, for example, the genes for seed color and seed coat are located on the same chromosome which would suggest interaction. But, since their locations are at remote sites and do not interact, Mendel's three-factor experiments support independent heritability.[16] Thus, as aptly summarized by Edwards, "the pea is in fact an excellent randomizer (Edwards 1986, p. 144)."

Mendel, however, was not so fortunate when it came to his experiments with Hieracium where there the anticipated laws of inheritance did not materialize because of the confounding complication of the asexual development of the embryo, apomixis. As ruefully noted by Mendel, "the [pea] progeny are variable and segregate according to a distinctive law," whereas "in Hieracium the direct opposite seems to reveal itself."[17] Mendell's success with *pisum* was thus a lucky accident since the principal confounding causes were auspiciously absent.

We turn now to a consideration of the proposal that Mendel, perhaps well-intentioned, somehow "cooked" his data in the hope that to do so would ferret out defective data points and retain only those worthy of consideration. The most promising approach along these lines begins with Edwards' analysis of the shape of the χ^2 values (when plotted against expectation) of the individual experiments.

[16]For more details on chromosomal locations and gene interaction see (Seidenfeld in Franklin et al. (2008), p. 250 n10) and Fairbanks and Rytting (2001, pp. 265–267). And for why the effects of genetic linkage would have been inaccessible to Mendel because of gene location and experimental design see (Fairbanks and Rytting in Franklin et al. (2008), pp. 288–292) and (Fairbanks in Franklin et al. (2008), pp. 308–309).

[17]Quoted and translated in Nogler (2006, p. 4). For a thorough review of Mendel's experiments with *Hieracium* see Bicknell et al. (2016).

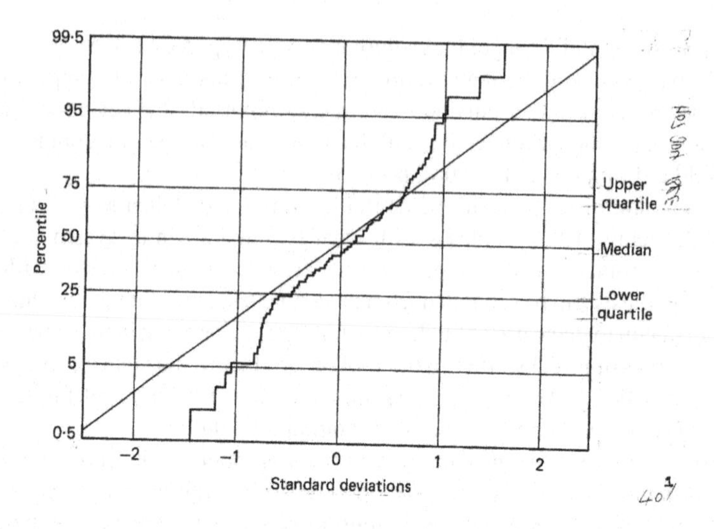

Fig. 5.1 The values of χ for the 69 segregations with undisputed expectations (the 2:1 ratio experiments are excluded) plotted on normal probability paper. *Source* Edwards (1986)

More precisely, in order to preserve the direction of the departure from expectation Edwards worked with χ rather than χ^2. Here the central insight is that what's revealing with respect to possible data manipulation by Mendel is the distribution of the 69 individual χ values (considered by Edwards) when plotted against expectation. In other words, that the shape of the curve, its signature as it were, would offer an important clue. And so it was, as can be readily seen by comparing this distribution as presented in Fig. 5.1 with the corresponding random normal deviates, Fig. 5.2, from tables provided in Lindley and Scott (1984).

The graph of Edward's results shows there's no significant deviations in the region between the 25th and 75th percentiles, but does show an excess of observations between the 5th and the 25th and the 75th and 95th percentiles, and a lack of observations between 0 and 5% and between 95 and 100%. In short, what's "too good to be true" is a feature of the tails of the distribution and not of the intermediate values. Thus, as concluded by Edwards:

> The inescapable conclusion is that some segregations beyond the outer 5-percentiles (approximately) have been systematically biased towards their expectations so as to fall between the 5-percentiles and the 25-percentiles (Edwards 1986, p. 159).

What this means is that, as shown by the shape of the χ curve, Mendel did not simply truncate data values but rather somehow moved (or as Edwards would later say "adjusted") some of the outer 5-percentile values into the inner space between the 5-percentiles and the 95-percentiles. Edwards, however, had no specific explanation as to how this systematic bias came about stating only:

> As to the precise method by which the data came to be adjusted, I would rather not speculate, *though it seems to me that any criticism of Mendel himself is quite unwarranted. Even if*

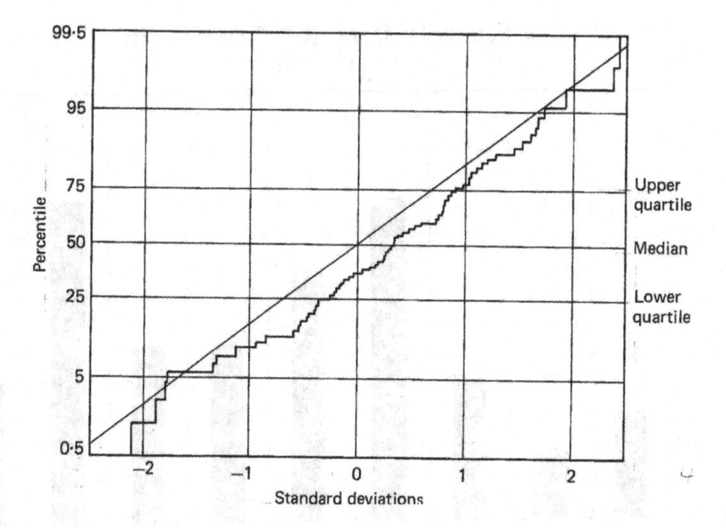

Fig. 5.2 The values of 69 simulated standard normal deviates plotted for comparison with Figure A1. *Source* Edwards (1986)

he were personally responsible for the biased counts his actions should not be judged by today's standards of data recording (Edwards 1986, p. 159, emphasis added).

Nevertheless, even if being judgmental is inappropriate, it would still be worthwhile to know exactly how Mendel might have managed such a sophisticated adjustment of his data. With a similar purpose in mind, Seidenfeld analyzed the *p*-values of the individual χ^2 tests. The results of this analysis, as exemplified in the histogram displayed in Fig. 5.3, were striking and revealing:[18]

1. There is a significant reduction in the left-tail of the Ps.
2. There is no significant departure from uniformity in the right tail of the Ps.
3. There is a significant concentration of the Ps about their median, i.e., about 0.50 (Seidenfeld in Franklin et al. 2008, p. 242).

Thus, Seidenfeld pointedly asked "[w]hat model of cheating, then, can the reader propose that replaces [such] extremely discrepant outcomes with ones clustered about the median of χ^2s", and then "challenge[d] the reader to try to adjust binomial data from samples sizes in Mendel's experiments, so that [these] three features appear in the resulting distribution of P-values from the (1 df) χ^2s" (Seidenfeld in Franklin et al. 2008, p. 242).

Ana Pires and João Branco took up the challenge and presented a well-developed "statistical model" which "may be seen as an approximation" for Mendel's procedure which "can clarify the controversy, that is, explain Fisher's conclusions without accusing Mendel (or any assistant) of deliberate fraud" (Pires and Branco 2010, p. 545, 557).

[18]Seidenfeld performed a similar analysis for Edwards' somewhat different partition of Mendel's data with essentially the same result. See (Seidenfeld in Franklin et al. 2008, p. 243).

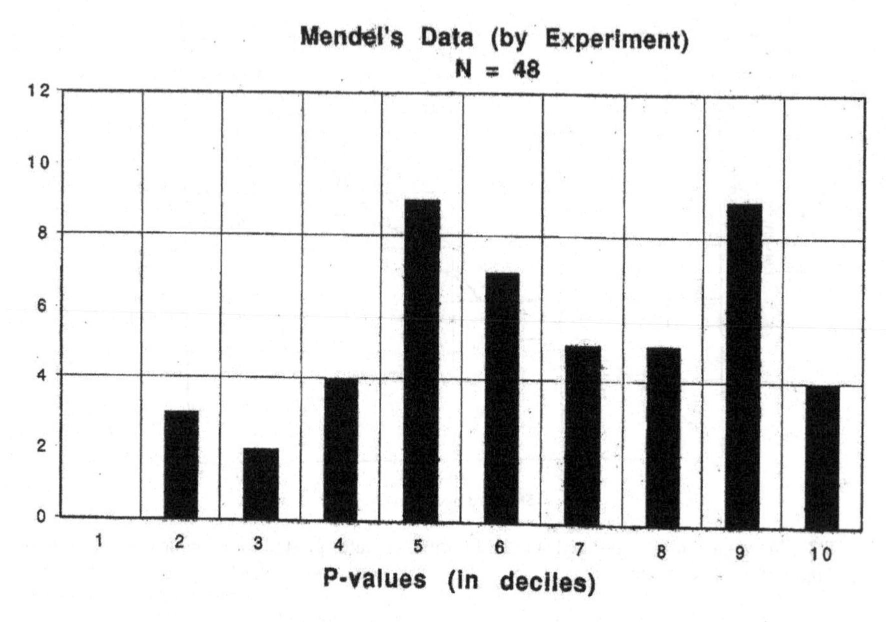

Fig. 5.3 Probability values by deciles for Mendel's data (by experiment). *Source* (Franklin et al. 2008)

The basic idea is simple enough:

Suppose Mendel has repeated some experiments, presumably those which deviate most from his theory, and reports only the best of the two (p. 553).[19]

Things, of course, get complicated when they convert this into a formal statistical theory.

> [A]n experiment is repeated whenever its p-value is smaller than α, where $0 \leq \alpha \leq 1$ is a parameter fixed by the experimenter, and then only the one with the largest p-value is reported (p. 553)

While obviously, though engagingly fanciful, Pires and Branco offer the following by way of defense:

> Note that this is just an idealized model on which to base our explanation. We are not suggesting that Mendel actually computed ***p***-values! p. 553, n4).[20]

Putting aside whatever qualms one might have about such idealized models, the principal virtue of the Pires and Branco explanation is that it does, in fact, precisely answer Seidenfeld's challenge. This is clearly and elegantly shown in Fig. 5.4 where

---

[19]By way of motivation Pires and Branco cite (Fairbanks and Rytting 2001, p. 743) ("We believe that the most likely explanation of the bias in Mendel's data is also the simplest. If Mendel selected for presentation a subset of his experiments that best represented his theories, χ^2 analysis of those experiments should display a bias.").

[20]Note, however, that Pires and Branco are assuming a deliberate action by Mendel.

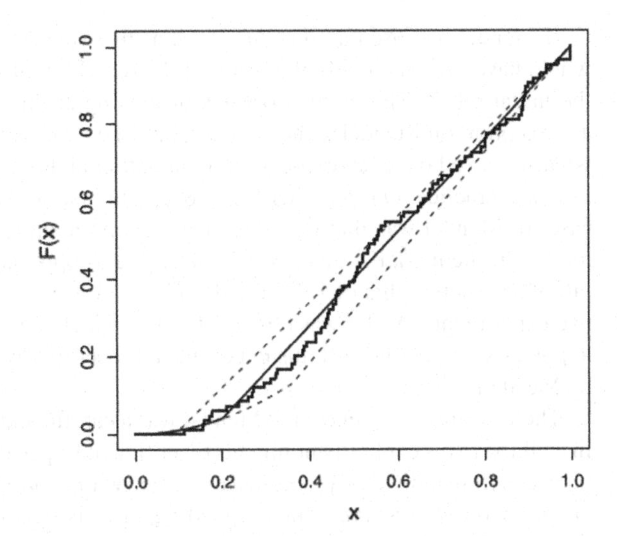

Fig. 5.4 Empirical cumulative distribution function of the p-values and fitted model (solid line: $\alpha = 0.201$; dashed lines: 90% confidence limits). *Source* Pires and Branco (2010)

the *p*-curve has exactly the features demanded by Seidenfeld's analysis.[21]

And if it's assumed that Mendel had some sort of surrogate (perhaps just his innate sense of statistics) for the α value parameter, then the model and the demonstration that it works as advertised does serve to give some indication of what it would have taken, in terms of an "adjusted" experimental procedure, for Mendel to have achieved the otherwise "too good to be true" experimental fit. In a similar vein, Jan Kalina refined and simplified the technical aspects of the Pires and Branco explanation and achieved a similar explanatory success. See Kalina (2014, pp. 95–97).

Still, while these ingenious explanatory efforts are thought provoking, there isn't convincing textual, or other, evidence that Mendel proceeded by once repeating a suspect experiment and then sticking with the better of the two. In fact, the best textual evidence that exists in this regard argues that Mendel did not follow this procedure. This comes from Mendel's expressly noted repeat of Experiment 5 of his F3 set of experiments on "The Second Generation [Bred] from the Hybrids" that we reviewed earlier. As the reader may recall, the first two experiments along with the sixth yielded "almost exactly the average ratio of 2–1." With respect to the third and fourth experiments "the ratio varies more or less, as was only to be expected in view of the smaller number of 100 trial plants." The fifth experiment, however, posed a problem because the plant ratio was 60–40 which, for Mendel was too far off the mark. Accordingly:

> Experiment 5, which shows the greatest departure, was repeated, and then, in lieu of the ratio of 60 and 40, that of 65 and 35 resulted. *The average ratio of 2 to 1 appears, therefore, as fixed with certainty* (Mendel in Bateson 1909, p. 89)

[21] It must be noted that Pires and Branco require a complicated determination ("a minimum distance estimator") of the value of the parameter α required to ensure this result. But for present purposes we can ignore this technical nicety. For the details see (Pires and Branco 2010, p. 554).

If Mendel had used a procedure such as that suggested by Pires and Branco he would have presented only the result of the repetition of Experiment 5 and omitted the initial result. There are two ways of looking at this. First, that the repeat was an exception on Mendel's part which is why he expressly noted it. In other words, what we have here is an expression of an outlier and not an implicit admission of a standard practice. On the other hand, one could argue that what distinguished this case for Mendel was that the "departure" from the anticipated value was not only too far off the mark but that it was exactly 3–2 as opposed to some smaller decimal difference such as that of 1.93–1 and 2.13–1 as was respectively the case in the first two experiments. And it was this exactitude—which may have suggested a contrary hypothesis—that rendered this experimental repetition worthy of express comment by Mendel.[22,23]

There is strong evidence that Mendel was a careful and scrupulous experimenter. In addition to his explicitly noting his repetition of Experiment 5 and his presentation of results from individual plants that disagreed with his model. Mendel also published his paper on *Hieracium,* which argued against his hypotheses. He also offered to send von Nageli samples of his seeds so that von Nageli could repeat Mendel's experiments. These are surely not the actions of someone committing fraud. Nor are they what one would expect if Mendel had indulged in an otherwise well-intended manipulation of the data.

We suggest that Mendel's experiments should have been enough in 1865 for not one, but at least two purposes. His experiments on pea plants not only suggested the laws of segregation and independent assortment, which are the basis of modern genetics, but also provided strong evidence for those laws. They were also important for the methodology they illustrated. As Fisher remarked

> The facts available in 1900 were at least sufficient to establish Mendel's contribution as one of the greatest experimental advances in the history of biology.. .. If we read his introduction literally we do not find him expressing the purpose of solving a great problem or reporting a resounding discovery. He represents his work rather as a contribution to the *methodology* of research into plant inheritance. (Fisher 1965, 2, 3, emphasis in original).

In sum, Fisher's appraisal—as we read him—was that Mendel's pea experiments were enough as they stood to justify the pursuit and further development of the

[22]Pires and Branco make a determined, though we think unconvincing, effort to marshal additional supporting evidence from Mendel's 1866 paper. See (Pires and Branco, 2010, pp. 556–557). For example, although the results from Experiment 7 are almost as far from Mendel's expectations as those of Experiment 5, he did not repeat that experiment. We should also note that Mendel included data from individual plants that disagreed strongly with his hypothesis.

[23]Here we mirror in part Seidenfeld's appraisal that "unless some alternative model with reduced variance, like the C-P model, can be justified, I see little hope of explaining away the Ps that are 'too good to be true'" (Seidenfeld 2008, 243). To further complicate the matter see Noel Ellis and Hofer (2019, p. 1), which considers the possibility that Mendel did not perform his pea experiments in the order presented in his 1865 paper, and that "two of his F3 progeny test experiments may have involved the same F2 population, and therefore that these data should not be treated as independent variables" and as a result Mendel's segregation ratios do not differ "remarkably from expectation." And to think that all of this could have been avoided if only Mendel's notebooks had not been burned at the time of his death (Fairbanks and Rytting 2001, p. 265).

methodology employed and its consequences for what ultimately became known as genetics.[24]

It is tempting to view Mendel's pea experiments as being on par with, for example, the Meselson-Stahl experiment which, as we have noted, was regarded as the "most beautiful experiment in biology." And indeed there have been many such appraisals made of Mendel's work holding it for example, to be "a supreme example of scientific experimentation" (Stern and Sherwood 1966, p. v) and "a model in respect of the order and lucidity with which the successive relevant facts are presented" (Fischer 1936, p. 119). To which should be added the many attributions of Mendel having been "the father of genetics." And if such appraisals be the case, then as a corollary one could justifiably claim that "once should have been enough" when Mendel's pea experiments first saw the light of day in 1865.

With respect to the accuracy and reliability of Mendel's results it is important to emphasize that the "replications" by de Vries, Correns, and von Tschermak were not undertaken in order to test the reproducibility of Mendel's results but rather had been independently conducted in the absence of any knowledge of Mendel's work and only later were realized to have in essence duplicated Mendel's results.[25] There were, however, replications of a different sort later performed with knowledge of Mendel's work. Without delving into the details, it is fair to say that the intent of these experimental endeavors was not so much to test Mendel but rather to elaborate and expand on his methodology. As William Bateson said of his own work:

> Though Mendel's results have been amply confirmed by Tschermak and Correns, [my] experiments with *Pisum* were made partly from a desire to witness the classical phenomena and partly in the hope of elucidating certain obscure points (Bateson and Sanders 1904, p. 55).

In a similar vein, A.D. Darbishire's aim was to further determine and examine the relationship between "the starch-grains of round and wrinkled peas" which was important because the difference "afforded an infallible test by which the real character of a pea with doubtful shape could be determined" (Darbishire 1907, p. 122).[26]

---

[24] As we indicated in an earlier chapter, this distinction was first proposed in Laudan (1977, pp. 100–114), questioned in Nickles (1981, pp. 100–108) and applied to certain episodes in particle physics in Franklin (1993). Here the central distinction is between *acceptance* (no loose ends so far or anticipated) and *pursuit* (more incoherence and loose ends than the competition but nevertheless worth pursuing). A close relative is Thomas Kuhn's much discussed concept of a *paradigm* which (as we understand it) is a stunning experimental or theoretical success (often limited in scope) that despite it success is far from perfect, requires extensive development and expansion in terms of scope, and thereby provides well motivated (and career enhancing) employment possibilities for scientific practitioners (Kuhn 1962, pp. 43–51).

[25] This of course leaves open the question whether de Vries, Correns, and von Tschermak would have thought that exact (or nearly exact) replication would have been called for if they had first known of Mendel's pea experiments, or whether as in the case of the replications by Darbishire and others that followed (see below) they would have pursued a different course of experimentation designed to further develop and expand upon Mendel's results.

[26] For an extensive review of extensions of species studied and experimental refinements see (White 1916, 1917).

One might have expected a call for a more focused replication in the sense of validation when Fisher announced his χ^2 results. But by then Mendel had already in large part won the day when it came to the development of genetics. Which rendered Fisher's analysis essentially moot, and which explains why it attracted little attention until its revival around the time of the 1965 Mendel Centenary.[27] Moot or not, Fisher went on to affirmatively assert that Mendel's "experimental researches [were] conclusive in their results, faultlessly lucid in presentation, and vital to the understanding not of one problem of current interest, but of many" (Fischer 1936, p. 139).

In sum, Fisher's appraisal—as we read him—was that Mendel's pea experiments were enough as they stood to justify the pursuit and further development of the methodology employed and its consequences for what ultimately became known as genetics. And once this development had successfully proceeded, one could in retrospect credit Mendel with having been the "father of genetics" and thereby confirm the soundness of the determinations that *once was enough* in 1900, and *should have been enough* in 1865. Understood this way, whatever defects that may have existed in Mendel's data collection were subsumed in the process of the further successful development and elaboration of Mendel's methodology and the corresponding laws of inheritance. Fisher, we suggest, was thinking along similar lines when he remarked:

> Only a succession of publications, the progressive building up of a corpus of scientific work, and the continuous iteration of all new options seem sufficient to bring a new discovery into general recognition. (Fischer 1936, p. 139)

One last methodological point about replication is in order. As we argued in our Introduction, what's necessary is not that an experiment be exactly (or nearly so) repeated but rather that there be evidence that *if* the experiment were exactly (or nearly so) repeated then the same result would be obtained. In Mendel's case this sort of evidence consisted of the successful development and elaboration of his work. In a nutshell, successful pursuit establishes in effect the functional equivalent of an actual replication of an experiment.

---

[27] One question that occurs to us is why did the revival of interest in Fisher's "too good to be true" argument blossom around the time of the 1965 Mendel Centenary and not earlier at the time of Mendel's 1950 "Golden Jubilee" (celebrating the rediscovery of Mendel in 1900). For a fascinating hint as to why this might have been see Wolfe (2012) who demonstrates that "[t]he Golden Jubilee may have primarily been 'about' Mendel, but it was just as much about demonstrating the practical achievements of Western science as a proxy for the superiority of the American way of life at the dawn of the Cold War" (p. 393). Given this, it seems to us that it would hardly have been in keeping with this purpose to resurrect Fisher's argument that all of this was based on Mendel's cooked data. Much to our disappointment there is no evidence to suggest that such a consideration was ever at issue. For the considerations (many of them serious and informative) that did play a role see Wolfe's fascinating article.

References

Bateson, W. 1909. *Mendel's principles of heredity*. Cambridge: Cambridge University Press.

Bateson, W., and E.R. Sanders. 1904. *Experimental studies in the physiology of heredity reports to the evolution committee of the royal society II*. London: Harrison and Sons, St. Martin's Lane.

Beadle, G.W. 1967. Mendelism, 1965. In *Heritage from Mendel*, ed. R.A. Brink, 335–350. Madison: University of Wisconsin Press.

Bicknell, R., A. Catanach, et al. 2016. Seeds of Diubt: Mendel's choice of *Hieracium* to study inheritance, a case of right plant, wrong trait. *Theoretical and Applied Genetics* 129: 2253–2266.

Correns, C. 1900. G. Mendels Regel uber das Verhalten der Nachkommenschat der Bastarde. *Berichte der Deutschen Botanischen Gesellschaft* 18: 158–168.

Darbishire, A.D. 1907. On the Result of Crossing Round with Wrinkled Peas, with Especial Reference to their Strach-grains." *Proceedings of the Royal Society (London), Seies B* **80**: 122-135.

de Vries, H. 1900. Das Spaltungsgesetz der Bastarde. *Berichte der Deutschen Botanischen Gesellschaft* 18: 83–90.

Edwards, A.W.F. 1986. Are Mendel's results really too close? *Biological Reviews* 61: 295–312.

Fairbanks, D.J., and B. Rytting. 2001. Mendelian controversies: A botanical and historical review. *American Journal of Botany* 88: 737–752.

Fisher, R.A. 1936. Has Mendel's work been rediscovered? *Annals of Science* 1: 115–137.

Fisher, R.A. 1965. Introductory notes on Mendel's paper. In *Experiments in plant hybridisation by Gregor Mendel, with commentary and assessment by Sir Ronald Fisher*, ed. J.H. Bennett, 1–6. Edinburgh: Oliver and Boyd.

Focke, W.O. 1881. *Die Pflanzen-mischlinge ein Beitrag zur Biologie der Gewächse*. Berlin.

Franklin, A. 1993. Discovery, pursuit, and justification. *Perspectives on Science* 1: 252–284.

Franklin, A., A.W.F. Edwards, et al. 2008. *Ending the Mendel-Fisher controversy*. Pittsburgh: University of Pittsburgh Press.

Kalina, J. 2014. Gregor Mendel, his experiments and their statistical evaluation. *Acta Musei Moraviae, Scientiae biologicae* 99: 87–99.

Kuhn, T. 1962. *The structure of scientific revolutions*. Chicago: University of Chicago Press.

Laudan, L. 1977. *Progress and its problems: Toward a theory of scientific growth*. Berkeley: University of California Press.

Lindley, D.V., and W.F. Scott. 1984. *New Cambridge elementary statistical tables*. Cambridge: Cambridge University Press.

Mawer, S. 2006. *Gregor Mendel: Planting the seeds of genetica*. New York: Abrams.

Mendel, G. 1870. Uber einige aus kunstlicher Befruchtung gewonnen Hieracium-Bastarde. *Verhandlungen des naturforschenden Vereines in Brunn* 8: 26–31.

Nickles, T. 1981. What is a problem that we may solve it? *Synthese* 47: 85–118.

Noel Ellis, T.H., J.M.I. Hofer, et al. 2019. Mendel's Pea Crosses: Varieties, traits, and statistics. *Hereditas* 156(Article 33): 1–11.

Nogler, G.A. 2006. The lesser-known Mendel: His experiments on hieracium. *Genetics* 172: 1–6.

Pires, A., and J. Branco. 2010. A statistical model to explain the Mendel-Fisher controversy. *Statistical Science* 25: 545–565.

Stern, C., and E.R. Sherwood (eds.). 1966. *The origin of genetics: A Mendel source book*. San Francisco, W.H: Freeman and Co.

Thoday, J.M. 1966. Mendel's work as an introduction to genetics. *Advancement of Science* 23: 120–124.

Tschermak, E.V. 1900. Ueber kunstliche Kreuzung bei Pisum Sativum. *Berichte der Deutschen Botanischen Gesellschaft* 18: 232–239.

White, O.E. 1916. Inheritance studies in Pisum. I. Inheritance of cotyledon color. *The American Naturalist* 50 (579): 530–547.

White, O.E. 1917. Inheritance studies in Pisum. II. The present state of knowledge of heredity and variation in peas. *The American Naturalist* 56 (7): 487–588.

Wolfe, A.F. 2012. The cold war context of the Golden Jubilee, or, why we think of Mendel as the Father of genetics. *Journal of the History of Biology* 45: 389–414.

Part II
Once Wasn't Enough, Later It Was

Chapter 6
Once Came Close to Being Enough: Electron Polarization and Parity Nonconservation

6.1 Introduction

Our focus so far has been on cases where once was enough in the sense that replication of data and result was not called for either historically or in retrospect as a matter of justifiably required scientific practice. By so doing our hope is that considering cases at that end of the replication spectrum would yield insights not otherwise gained. But to so focus runs the risk of making the role of experimentation appear too neat and tidy, and moreover one characterized by prompt resolution of whether once was enough.

To counterbalance such risk but at the same time retain something of our original focus we will in this chapter deal with a case where *once was almost but not quite enough* and because of that the resulting historical process was highly convoluted and contentious, with missed opportunities, and experiments left for dead only to be later resurrected—all capped by considerable irony. Add to this a gestation period of more than twenty-five years and what we've got stands in strong but not total contrast to our previous case studies.

6.2 Electron Polarization and the Cox Double Scattering Experiment

Richard Cox entitled his 1928 paper as "Apparent Evidence of Polarization in a Beam of β-Rays (Cox et al. 1928)."[1] The fact that the polarization involved had to do with electrons would have sounded decidedly odd a decade earlier. There were two significant developments that explain Cox's title and subject matter. First, in 1923 de Broglie had proposed that just as light exhibited both particle and wave

[1] Cox had two co-authors, Charles McIlwraith and Bernard Kurrelmeyer. Cox was the group leader and primary author. For convenience we will often refer to the Cox experiment.

A. Franklin and R. Laymon, *Once Can Be Enough*,
https://doi.org/10.1007/978-3-030-62565-8_6

characteristics, so should what we normally consider particles, exhibit wave characteristics (De Broglie 1923a, 1923b). This had been brilliantly confirmed in 1927 by Davisson and Germer in their experiment on the diffraction of electrons by crystals (Davisson and Germer 1927). The second development was prompted by the fact that the hydrogen spectrum gave twice as many levels as would otherwise be expected. As aptly later summarized by C.G. Darwin: "This doubling was explained in the old quantum theory by the idea of the spinning electron, a convenient term to describe the quality of possessing a definite angular momentum and magnetic moment as well as charge (Darwin 1930, p. 379)." Thus, Uhlenbeck and Goudsmit took advantage of the added degrees of freedom afforded by a spinning electron to explain the hitherto puzzling existence of such doubling (Uhlenbeck and Goudsmit 1926).

Cox, to his credit, recognized that given de Broglie's waves there was an *analogy* to be made between the spinning electron and the electric and magnetic fields of ordinary light—and moreover that this analogy could serve as the basis for an experimental test for the existence of electron polarization. As neatly expressed in the first few sentences of Cox's paper.

> The already classic experiment of Davisson and Germer in which the diffraction of electrons by a crystal shows the immediate experimental reality of the phase waves of deBroglie and Schrödinger suggested that it might be of interest to carry out with *a beam of electrons experiments analogous to optical experiments in polarization.* It was anticipated that the electron spin, postulated by Compton to explain the systematic curvature of the fog-tracks of β-rays, and recently so happily introduced in the theory of atomic spectra by Uhlenbeck and Goudsmit might appear in such an experiment as the *analogue of a transverse vector in the optical experiments.* This idea has lately been developed by Darwin (Cox et al. 1928, p. 544, footnotes omitted, emphasis added).

The general suggestion of such an analogy had been made by Darwin (in the reference cited by Cox), who proposed that:

> All these objections [to the combination of spinning electron and wave theory] are met at the outset if we take the analogy of light and assume that, just as there are two independent polarized components in a wave of light, so there are two independent components in the wave of an electron (Darwin 1927, p. 230).

Thus, the spinning electron could serve as the *surrogate* for the electric and magnetic fields of ordinary light and thereby provide the basis for a coherent concept of electron polarization. So just as optical polarization refers to the transformation of a prior uniform distribution of the electric and magnetic fields to a distribution less uniform, so too would electron polarization refer to the transformation of a prior uniform distribution of spin values to a distribution less uniform. Cox's problem then was to devise an experiment test would test for electron polarization. And here he hit upon the idea of using Charles Barkla's demonstration that X-rays could be polarized as the basis for an analogous test of electron polarization.

Looking for inspiration in Barkla's experiment was particularly appropriate because unlike ordinary light, X-rays are reflected diffusely and thus cannot be readily polarized by reflection or by well-behaved refraction. So too with beams of electrons. Thus, as stated by Cox:

> Since the equivalent wave-length in the wave mechanics of even slowly moving electrons is of the order of that of x-rays, it seemed preferable to attempt an experiment analogous to optical polarization by *scattering or absorption* rather than by reflection or double refraction. We have been *chiefly occupied* by an experiment *analogous* to that in which Barkla demonstrated the polarization of x-rays by double scattering (Cox et al. 1928, p. 545, emphasis added).

Barkla's problem was to devise an experiment that would confirm the theoretical prediction by J.J. Thompson that if X-rays produced by a cathode ray tube were made to impinge on a substance (known as the "radiator") they would produce a secondary X-ray that was plane polarized and perpendicular to the original X-ray. It was empirically determined that carbon in particular worked well as a radiator and produced a substantial and well-behaved secondary X-ray. See Barkla (1906).

Referring to Barkla's initially bewildering (almost comically so) diagram (here reproduced as Fig. 6.1) the basic idea was to place a cathode ray tube (K) and a corresponding carbon "radiator" (R_1) in a lead encased enclosure where the secondary presumably polarized X-ray beam could escape through an adjustable aperture (C_1 and S_1) and make its way through another adjustable aperture (C_2 and S_2) to a second radiator (R_2) where a tertiary X-ray beam would be produced. Two perpendicular opportunities were provided for the escape of the tertiary beam. First, horizontally through the 5 × 5 cm aperture C_3 in the vertical wall S_3 where it would meet up with the detector electroscope A_1. Second, in a *vertical* trajectory up to the *horizontally* placed detector A_2 (so that A_2 sits atop of R_2). In between R_2 and A_2 there's a 5 × 5 cm aperture—but where the identification of this aperture was mistakenly not indicated in the diagram but described in the text. Call it C_4. There's was also, as indicated, provision made for collecting X-rays that made their way through R_2.

With the apparatus in place, we now turn to the experiment. First, measurements were made of the intensities of the horizontal and vertical components of the tertiary

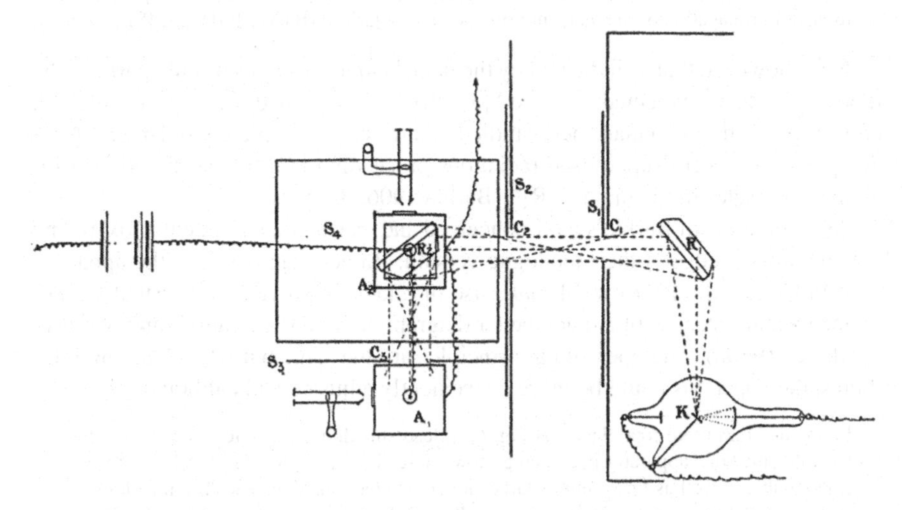

Fig. 6.1 Plan of apparatus, showing position of bulb giving maximum deflection of electroscope A_1 and minimum of electroscope A_2. *Source* Barkla (1906)

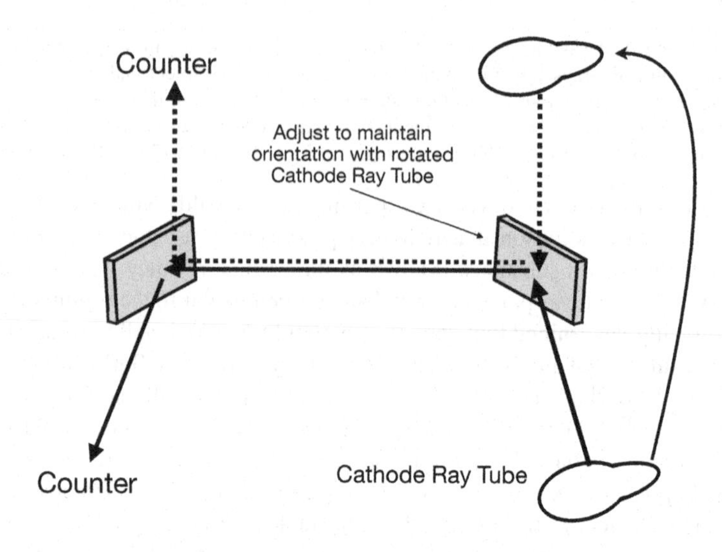

Fig. 6.2 Schematic of Barkla X-ray experiment showing tertiary beams of maximal intensity. *Source* Authors

radiation which were found to be respectively 6.3 and 1.9 in electroscope scale units. Second, in order to test for confounding apparatus asymmetries the cathode ray tube was rotated vertically and the experiment repeated where the respective intensities were found to be 1.95 and 5.85. See Fig. 6.2. Thus, as summarized by Barkla:

> It was found that the intensity of tertiary radiation reached a maximum when the directions of propagation of the primary and tertiary were parallel, and a minimum when they were at right angles, showing the secondary radiation proceeding from carbon in a direction perpendicular to that of propagation of the incident primary to be polarized (Barkla 1906, p. 248).

Note, however, that as indicated by the data the polarization was only partial. This though was to be expected because "[a]s the beams studied were of considerable cross-section, the secondary here studied could not be completely polarized, for at any point there were superposed radiations proceeding in different directions from all the corpuscles in the radiator R_1" (Barkla 1906, p. 253).

Armed then with Barkla's well-conceived and executed experiment, Cox and his collaborators produced their analogue version. Rather than describe the apparatus used in Cox et al. (1928) we'll make use of a later improved version that's easier to understand (because of a much better diagrammatic representation) and was used in Chase (1930b). The operational principles are the same but where, among other things, the apparatus could be more conveniently adjusted and calibrated.

> In this diagram [here reproduced as Fig. 6.3], lead shielding is depicted by double cross-hatching, the brass apparatus itself being shown in section by single cross-hatching. The two lead targets, as well as the radium source, are completely inked in. The sketch is drawn to scale. The radium source consists of a number of old radium emanation tubes containing radium D, E, etc., cut in half and bound in a bundle, open ends toward the upper target. The source is firmly attached to the shaft carrying the upper target, which can be rotated by the

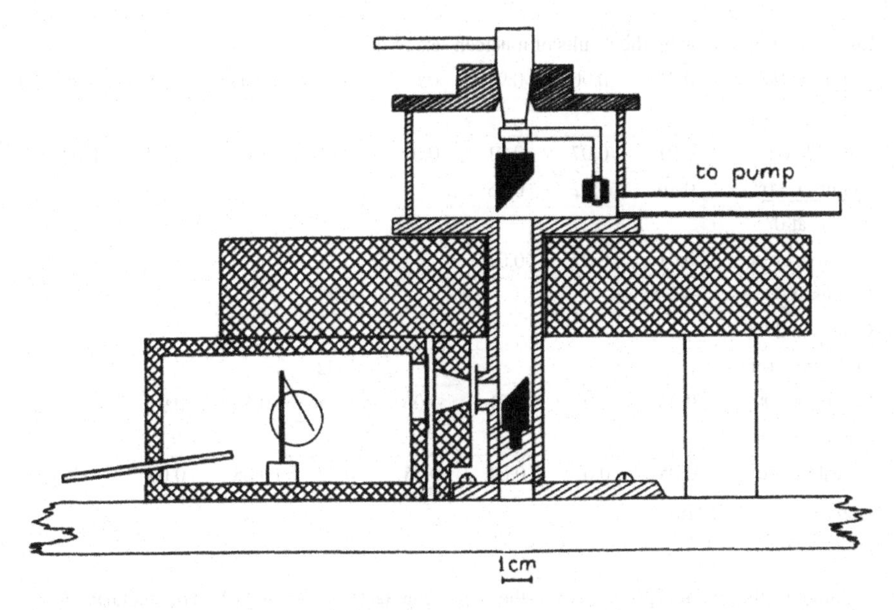

Fig. 6.3 Diagram of Chase apparatus at position 0°. *Source* Chase (1930b)

> handle shown at the top, the shaft passing through the cover in a ground joint. By this means the configuration of the electron path is altered (Chase 1930b, p. 1060).

The angular rotational position of the connected source and the upper lead target as shown in Fig. 6.3 is denoted as 0° where the configurations 90°, 180° and 270° correspond to clockwise rotations as seen from above (Chase 1930, p. 1062). As can be seen, the components closely match those employed by Barkla. So, whereas Barkla used two carbon radiators to polarize and analyze incoming X-rays, Chase used two lead targets to polarize and analyze the β-rays. Similarly, the intended experimental comparison was effectuated by respectively rotating the cathode ray tube or the radium source. Most importantly, the beam paths in both cases consisted of two perpendicular junctures. Finally, in both cases the sources were rotated, but Barkla did this to test for confounding apparatus asymmetries while Cox rotated the source in order to create a different path configuration and thus to test for revealing intensity differences.

The fly in the ointment of this otherwise well executed analogy of the Barkla apparatus was the inconsistency and limited operational lifetime of the Geiger counter used. This meant that extensive data was collected only for what was initially found to be the orientations of the radium source that gave the greatest difference in intensity outputs which was at the 90° and 270° locations. See Table 6.1 where the first three rows report the mean values of the early observations, and the penultimate row reports where later attention was focused on the comparison between the 90° and 270° orientations.

Table 6.1 Table showing the results of nineteen sets of data

Counts at 90º	0.76	0.90	0.94	0.87	0.98	1.03	1.03	0.91
Counts at 270°								
Probable error	0.01	0.07	0.01	0.02	0.01	0.03	0.02	0.02
Counts at 90º	0.76	0.62	0.65					
Counts at 0°								
Probable error	0.01	0.02	0.01					
Counts at 0º					0.87			
Counts at 270°								
Probable error					0.02			
Counts at 90º	0.95	0.99	1.01	1.06	1.05	0.55	0.91	
Counts at 270°								
Probable error	0.05	0.03	0.04	0.05	0.02	0.05	0.03	

Source Cox et al. (1928)

> Although the platinum points [used in the Geiger counter] were found the best of several types and materials that were tried, they are far from satisfactory. They usually give inconsistent results after an hour or two of use and have to be replaced. Moreover, the counts obtained with two different points do not agree. For this reason and on account of the uncertainty of the effect of the [to some degree unavoidable] γ-rays, it seemed inadvisable to attempt counts all around the circle. Attention was given instead to taking counts to test an early observation that fewer β-particles were recorded with the radium at 90° than at 270° (Cox et al. 1928, p. 546).

Probable errors were "computed in the usual way" (Cox et al. 1928, p. 547), and are indicated in the rows following the initial observations and the final set of observations. Presumably "the usual way" means that such errors were construed as a form of sampling error based on an assumed Gaussian distribution. If so, then these probable errors give no more than counting errors as in recording the Geiger meter counts or the later used electroscope displacements. One way to construe the experiment is as a test of three competing hypotheses: (1) the asymmetry values give a measure of the various confounding causes such as undetected asymmetries in the construction and alignment of the apparatus; (2) the asymmetry values give a measure of the polarization differences between the 90° and 270° orientations; or (3) some combination of (1) and (2). A sampling error calculation thus tells us only about the accuracy of the asymmetry values with respect to the first hypothesis, the second hypothesis, or the third hypothesis. Without additional information it does not distinguish between the three hypotheses.

With respect to such additional information there was the fact that the asymmetries between the 90° and 270° orientations fluctuated between positive and negative values (i.e., values greater or less than one) as well as values greater or less than the estimated error. This variation suggested some form of confounding error that varied *during the course of the experiment* as opposed to a fixed constructional or alignment difference.

After drawing attention to these fluctuations, Cox and his collaborators (henceforth denoted simply as "Cox") made the following response.

> It will be noted that of these results a large part indicate a marked asymmetry [between the 90° and 270° orientations]. The rest show no asymmetry beyond the order of the probable error. The wide divergence among the results calls for some explanation, and a suggestion to this end will be offered later. Meanwhile, a few remarks may be made on the qualitative evidence of asymmetry. Since the apparatus is symmetrical in design as between the two settings at 90° and 270° the source of the asymmetry must be looked for in an accidental asymmetry in construction or in *some asymmetry in the electron itself* (Cox et al. 1928, p. 547, emphasis added).

The methodology employed in this search for "an accidental asymmetry in construction" was to repeatedly readjust the suspected asymmetries and then determine whether there was a *consistent bias* introduced which would constitute evidence of such asymmetry. So, for example:

> The radium and the point in the counter were doubtless not exactly centered. But they were removed and replaced repeatedly in the course of the observations, and it seems unlikely that their accidental dislocations could be so *preponderantly in one direction* as are the observations (Cox et al. 1928, p. 547, emphasis added).

Similarly, for possible asymmetries in target orientation, the existence some residual magnetic field in the apparatus, and the possibility that the electron was polarized in passing through some material in the apparatus. See (Cox et al. 1928, pp. 547–548). Cox then went on to make a more general methodological point:

> It should be remarked of several of these suggested explanations of the observations that their acceptance would offer greater difficulties in accounting for the discrepancies among the different results than would the acceptance of the hypothesis that we have here a true polarization due to the double scattering of asymmetrical electrons. This latter hypothesis seems the most tenable at the present time (Cox et al. 1928, p. 547).

What Cox seems to be getting at is that while one might try to construct some sort of mathematical modeling of the confounding causes for the discrepancies, that would be difficult to do with any degree of certainty given the likely complex interactions and the limited data on the variational effect of the confounding causes. Better then at the least for the time being to rest content with the hypothesis that there is "a true polarization due to the double scattering of asymmetrical electrons." This at least provides a *target*, i.e., a *substantive result, for any attempted replication.*

With respect to the "wide divergence among the results" (i.e., the variation between positive and negative values as well as values less than or greater than the estimated error), Cox proposed the following.

> The discrepancies observed we ascribe tentatively to a selective action in the platinum points, whereby some points register only the slower β-particles. Observations in apparent agreement with this assumption have recently been made by N. Riehl. It is necessary to suppose further that the polarization is also selective, the effect being manifest only in the faster β-particles. In support of this it may be remarked that a few observations we have just made seem to show that asymmetry *is more consistently observed* when a piece of celluloid or cellophane is placed in front of the counter to stop the slower β-particles. Perhaps the simplest assumption here is that only β-particles which are scattered without loss of energy show polarization (p. 548, footnote omitted, emphasis added).

In sum, (1) manipulation of suspected structural and alignment asymmetries indicated that they would not act in a coordinated way such as to significantly affect the experimental values as the apparatus was deployed in different orientations, and (2) Geiger counter inconsistencies especially with regard to slower moving electrons was the likely cause of the inconsistencies in asymmetries values obtained for different measurements at the 90° and 270° orientations.

The reader may have noted that one thing that is missing so far is any account of the underlying mechanism that produces the claimed polarization. Some indication of how Cox—an experimentalist and not a theoretician—conceived the underlying causation is given at the very end of the paper where he suggests the following account:

> We have made no attempt at a theoretical treatment of double scattering beyond a consideration of the question whether the results here reported are of an asymmetry of higher order than what might be expected of a spinning electron. The following suggestion is then offered not at all as a theory of the phenomenon but merely as a remark on the geometry of the experiment. If it be supposed that the spin vector of a moving electron is always at right angles to its velocity vector, and that when the electron is scattered at right angles its new velocity vector has the direction of the vector product of its former velocity and spin vectors and its new spin vector has the direction of its former velocity vector, then the observations here described will be qualitatively accounted for (Cox et al. 1928, p. 548).

In other words, the prior coordination between velocity and spin gets systematically disturbed by the first polarizing interaction with the reflecting foil in a way that affects the initial uniform distribution of the spin vectors. But there's no specific account of how this could bring about the observed difference between the 270° and 90° orientations. In the meantime, a well-developed theoretical account had been developed by Neville Mott and would soon make its first published appearance, but for now we'll turn our attention to the improved version of the experiment conducted by Carl Chase who was then a graduate student working under the supervision of Professor Cox.

6.3 The Chase Double Scattering Experiment and the Confrontation with Mott's Theory of the Experiment

In an effort to find ways of improving the Cox experiment Chase embarked on a determination of the electron velocity sensitivity of the Geiger counter when different voltages were applied to the counter. The motivation here was based on his survey of the available experimental and theoretical literature which revealed that:

> All experiments with slow electrons have given negative results, while some of the experiments with fast electrons have shown evidences of polarization effects. As far as any theory is available, the prediction is that the spin vector of the electron may possibly show itself as the analog of a transverse vector in the electron waves, but *only when the electrons have high velocities* (Chase 1930a, p. 984, footnotes omitted, emphasis added).

Thus, any improvement to the Cox experiment was dependent on a more efficient production and capture of high-speed electrons as compared with their low speed cousins. Using a magnetic spectrograph Chase determined that more low speed electrons were counted when lower voltages were applied to the counter. This result was further confirmed by the simple expedient of comparing capture rates at a low voltage before and "a piece of paper 0.28 mm thick" was interposed between the electron beam and the counter. But because the Geiger counter became ever more unreliable as the voltage was increased (in order to increase the count of high speed electrons), Chase realized it would be best to move on to a specially prepared gold leaf electroscope which was expected to be both more reliable and more efficient in the registration of high speed electrons. In addition to introducing the electroscope, Chase improved the experiment by increasing path lengths, using a better radium source, and using lead shields to reduce gamma ray interference.

Chase reported the culmination of his efforts in a 1930 article in *Physical Review*. The improvements made it possible to reliably obtain data for the 0°/180° orientations as well as the 90°/270° orientations (see Table 6.2). With respect to the 90°/270° orientations, Chase obtained results that equate to a ratio of count at 90°/count as 270°

Table 6.2 Experimental results of Chase (relative counts)[a]

At 0°	At 90°	At 180°	Weight	
1.000	0.972	1.009	1.024	1
1.000	0.975	1.075	1.075	1
1.000	0.997	0.986	1.005	1
1.000	0.990	0.986	1.015	1
1.000	0.988	1.000	1.008	1
1.000	0.994	0.976	1.010	1
1.000	1.034	1.041	1.044	1
1.000		0.950		4
	1.000		1.030	3
	1.000		1.040	3
		1.000	1.020	2
1.000		0.933		1
	1.000		1.030	2
		1.000	0.969	2
1.000			1.003	1
1.000		1.037		2
	1.000	0.933		2
1.000 ± 0.003	0.993 ± 0.003	0.985 ± 0.000	1.021 ± 0.003	Weighted means

[a]Experimental error = 1%
Source Chase (1930b)

$= 0.973 \pm 0.004$.[2] Of particular importance is the fact that Chase's improvements led to an *elimination* of what Cox had earlier characterized as "the discrepancies among the different results." As summarized by Chase:

> The following can be said of the present experiments; the asymmetry between the counts at 90° and 270° is always observed, which was in no sense true before. Not only every single run, but even all the readings in every run, with few exceptions, show the effect (Chase 1930b, p. 1064).[3]

This meant that confounding causes that unpredictably varied as the experiment was being performed had been if not eliminated at least reduced to within the threshold of the error interval. In support of this conclusion Chase conducted the following experimental manipulation in order to demonstrate that the inconsistent performance of previously used the Geiger counter was the likely confounding cause that was responsible for the earlier discrepancies.

> As an interesting sort of check, the apparatus that had previously given a negative result was set up again; with the counters used as they were before, at lower voltages, the results were negative as before, but with high voltages on the counter, high enough to ruin the point within an hour or two, the effect [i.e., the asymmetry in output at the two orientations] was very likely to appear. Making no changes except in the voltage on the counter, the effect could be accentuated or suppressed (Chase 1930b, p. 1064, footnote excluded).

Chase also collected data for the 0°/180° orientations and obtained a ratio for the count at 180° and the count at 0° as being equal to 0.985 ± 0.005. But the noticeable discrepancies among the different runs made this result suspect.

> It will also be noted that *in the average* more electrons are counted at 0° than at 180°, though *this effect is not so consistently observed* as that between the other two positions. [This average value of 0.985] must therefore be regarded as *the least reliable of the set* (Chase 1930b, p. 1064, emphasis added).

Keeping Chase's many improvements in mind the question arises whether once was enough to cement the result for the 90°/270° asymmetry, and to justify the pursuit of further improvements with regard to the result for the 0°/180° asymmetry.[4] Such questions, however, were stopped in their tracks with the appearance of Mott's *theoretical* analysis based on the Dirac equation of electron polarization. It must be keep in mind that the path from the Dirac equation to the tangible reality of electron

---

[2]Here we have simply taken the ratio of the final values for the 90° and 270° orientations, i.e., 0.993/1.021, and multiplied it by 0.005 (Chase' claimed overall experimental error of one percent) to get 0.005 as the (rounded) error interval. We made a similar determination for the 0° and 180° orientations.

[3]Each run consisted of several individual readings, and where the mean value of the readings was reported as the value of the run in Table 6.2.

[4]With respect to possible improvements for what Chase considered to be "several unsatisfactory things about this experiment" see (Chase 1930b, pp. 1064–1065). These included the need for better control of interfering gamma-radiation, diminished electron velocities, and tertiary production of electrons.

polarization had to be mediated by a great many assumptions, idealizations, approximations and massively complex computations.[5] According to Mott, the bottom line of this dauntingly difficult process was that:

> The greatest asymmetry, therefore, will be found in the directions TM, TM', in the plane of the paper [i.e., in the 0°/180° orientations]. In the plane through T_1T_2 perpendicular to the plane of the paper [i.e., in the 90°/270° orientations], the scattering *is* symmetrical about T_1T_2. It was in this plane that asymmetry was looked for and the asymmetry found by them must be due to some other cause (Mott 1929, p. 431 emphasis in original).[6]

Chase cannot have been pleased. Especially since his best result was obtained for the 90°/270° orientations, while his result for the 0°/180° orientations was admittedly "the least reliable of the set." Thus, Chase's results were close to being the exact opposite of Mott's theoretical predictions. Chase's response to Mott was predictably cool.

> Mott has predicted on the basis of the Dirac electron theory that this experiment should show a polarization effect in that an asymmetry should be observable between the two positions 0° and 180°. This effect is supposed to be more definite for fast electrons and heavy scattering materials. He predicts equal counts for the positions 90° and 270° (Chase 1930b, 1064).

Chase then went on to defend his "least reliable" result dealing with the 0°/180° asymmetry.

> A previous [1929] experiment of the writer [i.e. Chase] showed an effect of this nature, more electrons being counted in the 0° position than in the opposite position by 4 percent. Nothing was claimed for this result, because the targets were so close together that it seemed plausible that the electrons would find it easier to get through the apparatus in the 0° position. The targets are now farther apart, since fewer electrons per second are permissible with the increased sensitivity of the electroscope, and beam divergence is much less than before. The previous objection is, therefore, no longer valid. That the effect is not due to the penetration of gamma-rays is shown by the fact that the presence or absence of the lead shield between apparatus and electroscope had little if any effect. A piece of paper held in the path of the rays, in front of the electroscope, cut down this asymmetry to a very small figure (Chase 1930b, p. 1064).

Presumably the purpose of the rehabilitation was to show that he and Mott were in at least partial agreement on the 0°/180° asymmetry. Chase's defense of what he had claimed were his superior results for the 90°/270° asymmetry was indirect and consisted of a commentary on an "similar" experiment performed by Emil Rupp which apparently yielded the 0°/180° asymmetry but not the 90°/270° asymmetry. Rupp's zero result for the 90°/270° asymmetry, however, was claimed by Chase

---

[5]Included in these assumptions was that there be single, large angle scattering of high velocity electrons from a high Z (proton number) nucleus. In a later paper Mott relaxed the Z value approximation which was not valid for the gold leaves used by experimenters (Mott 1932).

[6]See the Appendix to this chapter for a summary of how these results were obtained and for more details on the role played by electron spin in the polarization experiments. For interested readers, one thing to note is the relative simplicity of Mott's argument for there not being a 90°/270° asymmetry as opposed to the more complicated argument for what constituted the 0°/180° asymmetry.

to be "in accordance with our contention concerning the electron speeds necessary to observe this effect" (Chase 1930b, p. 1064).[7] Moreover, Chase noted that in Rupp's experiment "instead of scattering the electrons at right angles, the rays were reflected at grazing incidence" which meant that Mott's analysis was not applicable because "[i]t has not been shown that Mott's prediction applies directly to reflection at grazing incidence" (Chase 1930b, p. 1064). And to end this first round of what was to become an ongoing battle between and among experimentalists and theoreticians Chase promised that:

> [My] next paper will deal with scattering at right angles, but the electrons will be those which have been scattered by *transmission through thin foils*, in contrast to the present case in which diffusely reflected electrons were used (Chase 1930b, p. 1065, emphasis added).

We have emphasized this promised change to scatter by transmission because it would ultimately lead to a resolution of what, as we shall see below, would become an experimental and theoretical impasse. The promised transmission experiments, however, were either waylaid or unsuccessful and no such results were reported by Cox or Chase until the early 1940s when, as we shall see, they had dramatic effect.

6.4 The Ten-Year Impasse Between Experiment and Theory

Now it might be thought that even given Mott's objections once was enough for Chase's 1930 report to at least have justified further pursuit on electron polarization and double scattering experiments. And so it did but only in the following complex way. What ultimately set the agenda for pursuit was a succession, with only few exceptions, of experimentally determined *null results* for the 0°/180° asymmetry. This, of course, made for a direct conflict between these experimental results and Mott's theoretical prediction. This in turn led to yet more experiments and various attempts to modify or limit Mott's application of the Dirac equation. What got lost in all this was Chase and Cox's favored result for the 90°/270° asymmetry. That result was by and large ignored. While the reasons for this lack of interest were not to our knowledge clearly articulated, it may have been at least in part because of Mott's more easily understood symmetry argument regarding the 90°/270° orientations that predicted a null result.[8] If so, then Mott's *theory in effect disconfirmed the experiment and not the other way round.* The focus then was to be on the more complicated part of Mott's theory (the 0°/180° asymmetry) that was clearly at odds with an increasingly large number of contrary null results.

Because this period of uncertainty lasting between 1930 and the early 1940s has been well reviewed in great detail elsewhere[9] and is in any case rather convoluted

[7]Here Chase was implicitly relying on the results reported in (Chase 1930a) which he had briefly reviewed earlier at (p. 1063).

[8]See the Appendix to this chapter for the details on the relative simplicity of this determination.

[9]See Franklin (1979, 232–245) and Darrigol (1984, 52–79).

in its evolution, we'll give here a just a few representative samples of the frustration and uncertainty involved. So, for example, after reviewing the confused experimental situation G.O. Langstroth offered some reasons why Mott's analysis did not apply.

> In view of the fact that practical conditions may be immensely more complicated than those of Mott's theory, it is not surprising that it does not furnish a guide, even in a qualitative way, to all of the above experiments. This may be due to (a) the fact that a large proportion of the beam scattered from a thick target consists of electrons which have undergone more than one collision, (b) the insufficiency of the theoretical model, (c) the inclusion of extraneous effects in the experimental results (Langstroth 1932 pp. 566–567).

E.G. Dymond thought the fault lay on the theoretical side, in particular with the Dirac equation itself. Thus, having considered and rejected a range of possible reasons for the discrepancy between theory and experiment, he concluded:

> We are driven to the conclusion that the theoretical results are wrong. There is no reason to believe that the work of Mott is incorrect;. .. It seems not improbable, therefore, that the divergence of theory from experiment has a more deep-seated cause, and that the Dirac wave equation needs modification in order to account successfully for the absence of polarization (Dymond 1934, p. 666).

In addition, Dymond (1932) found difficulties with his own experiment. He had initially reported a positive result that was five times smaller than the theoretical prediction. He later found a considerable experimental asymmetry in his apparatus and repudiated that result in a 1934 paper (Dymond 1934).

During this period the only positive results reported, other than those of Chase, were a series of experiments by Rupp (1929, 1930, 1931, 1932a, 1932b, 1932c, 1934, Rupp and Szilard 1931). Rupp's results were fraudulent. In 1935 he published a note (Rupp 1935) in which he withdrew the results of several, but not all, of his experiments on electron polarization. This paper included a note from a doctor stating that Rupp had suffered from a mental illness which did not allow him to distinguish between fantasy and reality.[10] Rupp's work merely complicated an already complex situation.

Similarly, George Thompson concluded that at bottom the Dirac equation was to blame.

> Most of Mott's work has been, I understand, checked independently; but unless there is some error in the calculations, we are driven to the conclusion that Dirac's theory cannot be applied to the problem of scattering by heavy atoms. It seems very unlikely that the presence of the outer electrons can make much difference to the scattering, which must be almost entirely nuclear. .. All the conditions of Mott's calculations were fulfilled in these experiments, but it seems possible that Dirac's theory doesn't hold in the intense field close to a gold nucleus where, of course, most of the scattering takes place (Thomson 1934, pp. 1070–1071).

H. Richter for his part placed the blame with Mott's analysis and accordingly remained agnostic as to whether the fault law deeper with the Dirac equation.

> Despite all the favorable conditions of the experiment, however, no sign of the Mott effect could be observed. *With this experimental finding, Mott's theory of the double scattering of*

[10]For a detailed history of Rupp's work see Darrigol (1984).

> *electrons from the atomic nucleus can no longer be maintained.* It cannot be decided here how much Dirac's theory of electron spin, which is at the basis of Mott's theory, and its other applications are implicated through this denial of Mott's theory (Richter 1937, p. 554).

To further muddy the waters, in 1939 K. Kikuchi reported results which seemed to be in agreement with "Mott's theory" (i.e., the combination of the Dirac equation and Mott's analysis). Nevertheless, he qualified his endorsement by noting:

> Of course, it is quite beyond the range of experimental technique to realize perfectly all conditions from which Mott's theory is originated. In fact, the thickness of the gold foil used may be too large to be applicable for the theory (Kikuchi 1940, p. 524).

It is fair to say that in early 1940 the situation with respect to Mott's prediction of "the greatest asymmetry" at the 0°/180° orientation was as follows. Although some experimentalists had observed such an asymmetry the preponderance of evidence was that it had not been observed. Various theoretical attempts had been made to resolve the apparent discrepancy between theory and experiment with no success. It is therefore the ultimate irony that the resolution of the impasse with regard to the 0°/180° asymmetry came about due to new investigations that Chase, Cox, and their collaborators first reported in 1940. However, it would take a much deeper theoretical advance—the nonconservation of parity—in order to resurrect Chase's 90°/270° orientation result from the long gone and forgotten.

6.5 Transmission Efficiency and the Resolution of the Impasse

The resolution begins with the publication in 1940 by Chase and Cox of a paper that reported their investigation on the *single* scattering of 50 keV electrons from aluminum. Their purpose was to directly test Mott's equation for "the fraction [of the incoming electrons] scattered into a small solid angle Ω around a direction making an angle θ with the direction of incidence (Chase and Cox 1940, p. 243)." This equation was significant because of the central role it played in Mott's more general analysis of the polarization process employed in the original experiments by Chase and Cox and their many progeny.

> The results of several experiments have indicated exceptions to this equation and to the closely related prediction of an asymmetry in double scattering, and it has been suggested that they show an actual invalidity of the Dirac equations in a range of electron speeds where it is hardly to be expected on other evidence. On the other hand, the experiments offer difficulties in technique, and not all the experimental results are in mutual agreement. It seems desirable therefore to have more observation on the subject (Chase and Cox 1940, p. 243).

During the course of their experimental investigation Chase and Cox came across a hitherto unnoticed difference in the intensities in reflected and transmitted electron beams.

> It appeared in a comparison of the intensities of the beams scattered at 90° on the two sides of the foil, the side on which the beam was incident and the opposite side. Single nuclear scattering should be equally intense on both sides. But the observed scattering was consistently more intense on the side on which the beam was incident (p. 246).

This difference was potentially significant because the previous experiments which gave no polarization effects, most notably those of Dymond and Richter, were reflection experiments where the polarization effect would be expected to be considerably smaller than if they had been conducted using transmitted electron beams. Cox and Chase, however, did not explore this possibility of an effect on polarization, but went on to show that:

> With allowance made for this asymmetry [between reflected and transmitted intensities] in comparing the scattering at angles above and below 90°, relative intensities at angles from 30° to 120° were within 5 percent of those predicted by Mott (Case and Cox 1940, p. 243).

So, while Mott's scattering equation was vindicated (to within 5%), this left open the quantitative significance of the transmission-reflection asymmetry for the polarization experiments. As an initial step in this direction Cox and G. Goertzel calculated the probability that there would be more (small angle) multiple collisions within the scattering foil when there was reflection as opposed to transmission. The importance of this difference is that what's now known as *plural scattering* results in less polarization (Goertzel and Cox 1943). Applying their analysis to Dymond's experiment which apparently showed a null result, Cox and Goertzel tentatively concluded that

> half or more of the scattering in Dymond's experiment was plural in spite of the precautions taken against it. The polarization in this event would have been much reduced, and *it seems possible* that it may have been so much diminished as to escape detection (Cox and Goertzel 1943, p. 40, emphasis added).

At about the same time, as reported in Shull (1942), and more extensively in Shull et al. (1943), a variant of the Chase polarization experiment was conducted in order to more precisely determine the connection between the reflection-transmission difference and the observed polarization asymmetry. The relevant differences in apparatus and procedure were as follows. First, counters were placed at *both* sides of the second (analyzer) foil so as to respectively and simultaneously count transmitted and reflected electrons. See Fig. 6.4. Second, both the polarizer and the analyzer foils could be rotated so as to favor either transmission or reflection.

In the first part of the experiment, the polarizing foil was set to favor transmission. See Fig. 6.5. When adding in the rotation of the electron source, there were no less than eight different configurations. See Fig. 6.6. Data were collected for *all configurations* on the basis of which (1) the "polarization ratio," that is, the number of electrons scattered to the 180° position divided by the number scattered to the 0° position was determined for gold foils to be 1.08 ± 0.01; and (2) the "reflection-transmission asymmetry," that is, the number of electrons scattered to the side of incidence of the foil divided by the number scattered at the 180° position divided by the number at the same angles through the transmission side of the foil, was determined for gold foils to be 1.55 ± 0.0.15 and for aluminum foils 2.12 ± 0.035 (Shull et al. 1943, p. 34).

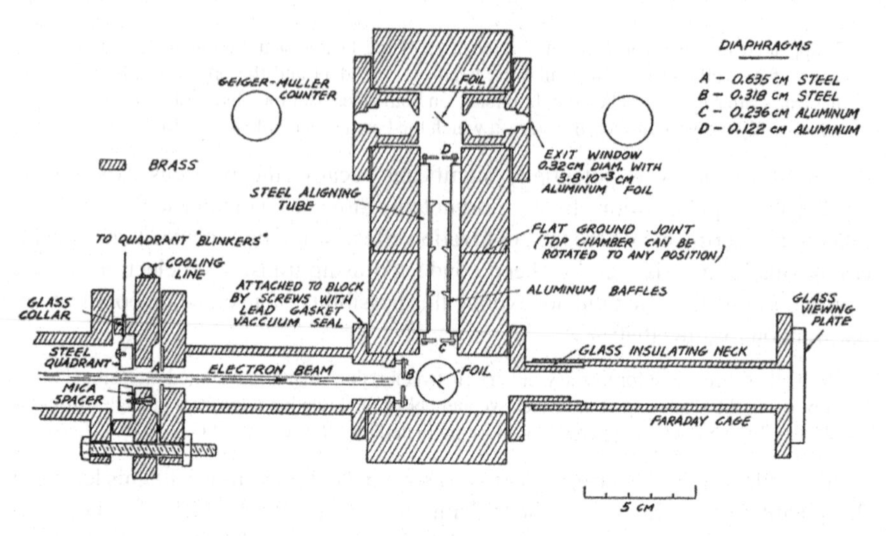

Fig. 6.4 Scattering chamber assembly. *Source* Shull et al. (1943)

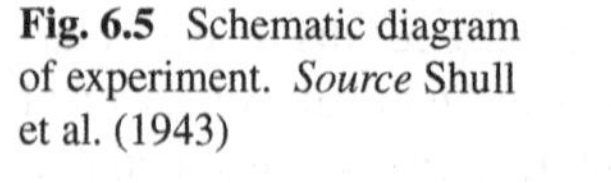
Fig. 6.5 Schematic diagram of experiment. *Source* Shull et al. (1943)

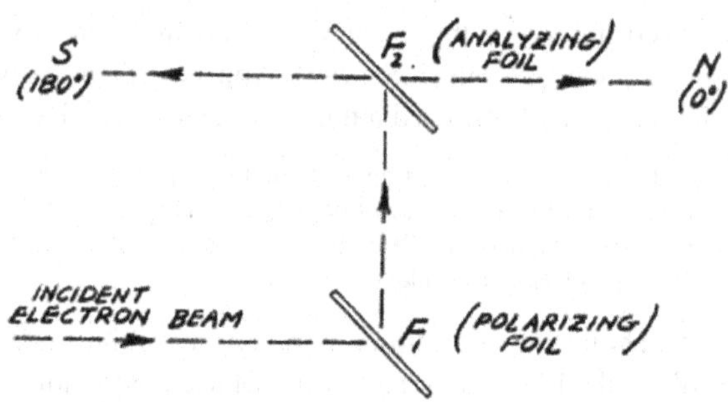

Now the reader may well wonder why were eight different apparatus configurations used when only two would have been enough? Answer: Because it provided for four *independent* determinations of the quantities involved. For the details see (Cox and Goertzel 1943, pp. 33–35). And because four independent and closely consistent determinations were made the possibility of an overriding systemic distortion was thereby more or less eliminated. Thus, by way of cautious understatement, it was commented that:

> This large [reflection-transmission] asymmetry is interesting in that past theories of particle scattering by thin foils have overlooked the possibility of such an effect. Goertzel and Cox [as described above] show that an effect of this kind may be caused by a type of plural scattering which consists in the combination of two deflections of the same order of magnitude.. .. If plural scattering is responsible for this effect, it is possible that a polarization experiment similar to the present one might be seriously affected by it because of the *depolarization which accompanies plural scattering*. There is the possibility then that a reflection polarization experiment (in which only "reflected" electrons are studied) will yield an asymmetry different from a transmission experiment (in which only "transmitted" electrons are studied). … *Since*

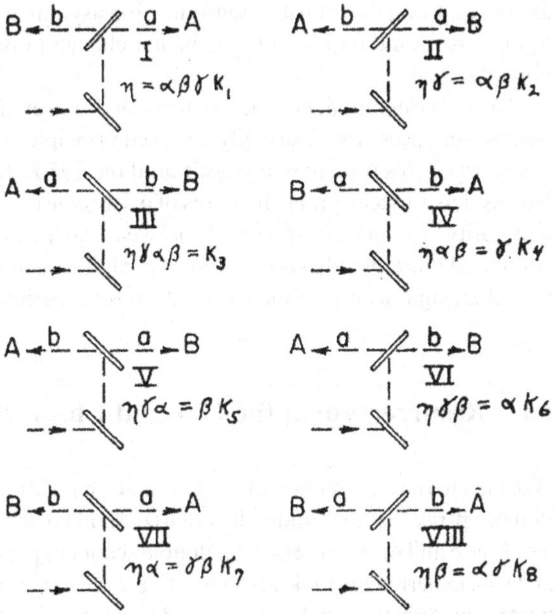

Fig. 6.6 Diagram which shows the eight different orientations that are possible with the analyzing foil set in the transmission orientation. *A* and *B* represent the two Geiger-Müller counters while *a* and *b* refer to the two exit faces of the analyzing chamber. *Source* Shull et al. (1943)

> *both "transmitted" and "reflected" electrons were utilized in calculating the polarization [ratio] value [of 1.08 ± 0.01], the result obtained is intermediate between that of a pure transmission experiment and that of a pure reflection experiment* (Shull et al. 1943, p. 34, emphasis added).

To obtain something of an approximation to a comparison of a "pure reflection experiment" to a "pure transmission experiment," the experiment was repeated for all configurations but this time setting the polarizing foil in a way that favored reflection. Using this additional data it was determined that the transmission asymmetry when the polarizing foil was set to favor transmission was "five times" greater than when the polarizing foil was set to favor reflection. Thus:

> The data presented above indicate a positive polarization effect which can be resolved into a transmission polarization asymmetry and a relatively small reflection polarization asymmetry (Shull et al. 1943, p. 34).

Armed with this result (so conscientiously achieved) all that remained to be done in order to more or less conclusively establish the 0°/180° asymmetry was to demonstrate that the reported null results of Dymond and Richter could be explained away on the basis of the results that Shull, Chase and Meyers had established. This was readily accomplished in the concluding section of the paper.

Thus here, at last, once was enough to establish the polarization of electrons and to justify the pursuit of increasingly more accurate experimental and theoretical determinations. For a comprehensive review of these further efforts see Tolhoek (1956). There remained, however, the 90°/270° anomaly which was still forgotten—apparently even by Chase and Cox since they made no mention of it in the papers discussed above regarding the transmission-reflection asymmetry. Bringing that anomaly back to the fore would require a fundamental change in theory, namely, that *parity* was not conserved.

Parity conservation, i.e., the existence and conservation of left-right (space reflection) symmetries, was a strongly believed principle. All known physical laws at the time obeyed this principle. It wasn't until the 1950s that another problem in physics led physicists to reexamine the status of parity conservation. After theorists suggested that parity was not conserved in the weak interactions and experiments strongly confirmed their hypothesis (see Chap. 7) physicists realized, at least retrospectively, the actual significance of the Cox and Chase experiments.

6.6 Resurrection of the Cox and Chase 90°/270° Anomaly

What led to the resurrection of the Cox and Chase 90°/270° anomaly was the realization that β-rays are longitudinally polarized and not unpolarized—as assumed by Cox and Chase and everyone else! All double scatter experiments after the original Chase and Cox experiments took advantage of the development of accelerators in which the beam size, direction, and energy could be controlled precisely—and which crucially produced unpolarized beams of electrons. Which meant, as correctly predicted by Mott, that there would be no asymmetry at the 90°/270° orientations. No physicist of the time thought that the difference between such artificially accelerated electrons and those from nuclear beta decay was of any significance. And this included Cox who later ruefully recalled that:

> For some years a small group of us at N.Y.U. continued experiments in the scattering and diffraction of electrons. But, as well as I can remember, most of our experiments were not with β-rays but with artificially accelerated electrons. Although the title of our first paper was "Apparent Evidence of Polarization in a Beam of β-rays", I did not suppose, and I do not think the others did, that β-rays were polarized on emission. I thought of the targets as having the same effect on any beam of electrons at a given speed, polarizing at the first target, analyzing at the second. Consequently, I did not think of the change from a radioactive source to an accelerating tube as a radical change in my field of research (Cox 1978, personal communication to Franklin).

Similarly, Mott admitted not to having realized the possibility that β-rays were longitudinally polarized which would make a difference in the polarization asymmetry detected in double scattering experiments.

> The extraordinary thing is that I had no idea that early attempts, unsuccessful, to find the effect I predicted were due to the nature of the β -ray source. I suppose that when I went to Bristol in 1933, I got so absorbed in solid state that I have paid very little attention to this sort of thing. Also I am not very proud of that particular paper.. .. I made some numerical mistakes in the first version which had to be corrected. However, I remember that Pauli read it, liked it, and offered me a place in his group at Zurich which regrettably I could not

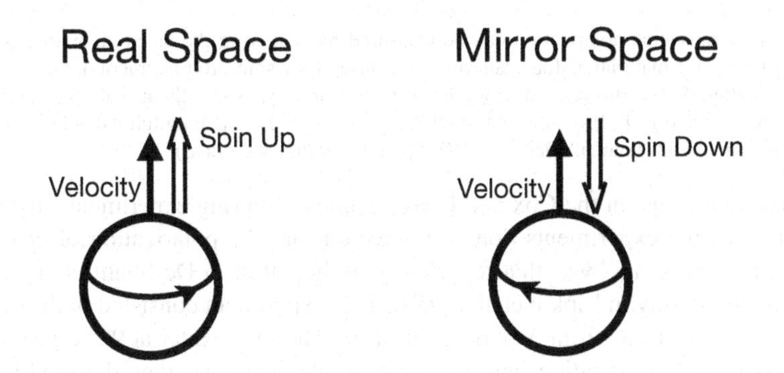

Fig. 6.7 Electron mirror image counterparts. *Source* Authors

take up, having already accepted something else (Mott, 1979 personal communication to Franklin).[11]

The realization that β-rays might be longitudinally polarized was an indirect result of the discovery of the parity nonconservation. This because if β-rays were longitudinally polarized that by itself would constitute an instance of parity nonconservation. To see why this is so we'll begin with the concept of a mirror image. Figure 6.7 shows an electron with spin and velocity in the same direction (here construed as a small rotating magnet) and its mirror image, where a mirror image is exactly what you would expect it to be, namely, the transformation defined by ($x \rightarrow -x$, $y \rightarrow -y$, $z \rightarrow -z$). The effect of the transformation is to change the spin from clockwise to counterclockwise, i.e., from spin up to spin down. Everything else remains the same, including in this case, the velocity vector. As will be explained more completely in the next chapter, parity conservation requires that all electrons and their mirror images be equally present. But if there is *longitudinal* polarization (i.e. *all* electrons in a beam are as depicted in Fig. 6.7) this requirement is not satisfied because the number of spin up electrons will not equal to the number of spin down electrons. In canvasing experimental methods for discovering parity nonconservation, Lee and Yang recognized the possibilities afforded by β-decay and accordingly suggested the measurement of momentum and polarization of the electrons emitted in such decay.

> Another interesting experiment is to measure the momentum and polarization of the electron emitted in a β decay. A polarization of the electron results only if parity is not conserved.... The polarization in such a case will be along the direction of the momentum of the electron (Lee and Yang 1957, p. 1674).

This effect was first demonstrated by Frauenfelder and his collaborators. Briefly this experiment consists of transforming the longitudinal polarization of the electrons to a transverse polarization and then detecting the transverse polarization by a left–right asymmetry in the scattering of the electrons.

> The observation of the expected longitudinal polarization is difficult. However, by means of an electrostatic deflector, the longitudinal polarization can be transformed into a transverse

[11]Thus, Pauli apparently missed the possibility as well.

> one. The transverse polarization can be measured by scattering the electrons with a thin foil of a high-Z material. (Mott scattering.) Because of the spin-orbit interaction, the elastically scattered electrons show a strong left-right asymmetry, especially at scattering angles between 90° and 150°. From this measurable asymmetry the initial longitudinal polarization can be calculated (Frauenfelder et al. 1957, p. 386, footnotes omitted).

Given our focus on the Cox and Chase double scattering experiments, the most important of the experiments done on detection of "the polarization of electrons emitted in beta decay" was that reported soon thereafter in De Shalit et al. (1957) and less extensively in Lipkin et al. (1958). This experiment consisted of the double scattering of electrons from two metal surfaces. The first scatter at 90° transformed the longitudinal polarization into a transverse one. This was then detected by the left-right asymmetry in the second scattering. See Fig. 6.8. The experimenters gave a helpful qualitative explanation of the operation of the experiment.

> To follow qualitatively the behavior of an electron in this setup, consider (Fig. 2) [here reproduced as Fig. 6.9] a nonrelativistic electron (its *magnetic moment* is represented by the heavy arrow) scattered by 90° at σ_1. Since the electron is nonrelativistic, only its momentum is affected by the first scattering, but its spin (and therefore its magnetic moment) retains

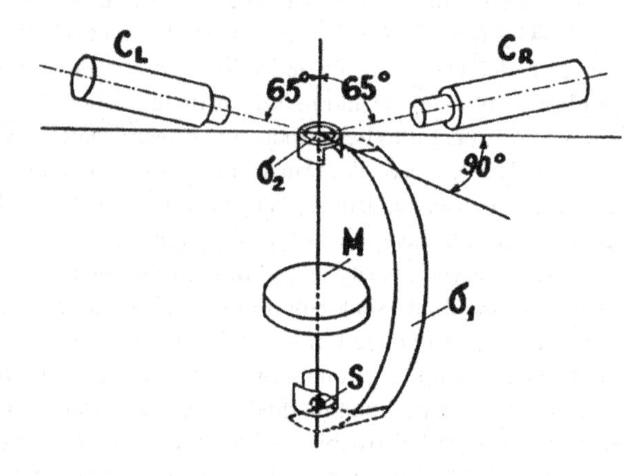

Fig. 6.8 Sketch of the setup for double scattering. *S* is the source; σ_1, an aluminum foil; σ_2, gold foil; *M*, lead to prevent electrons from reaching σ_2 directly from *S*. *Source* De Shalit et al. (1957)

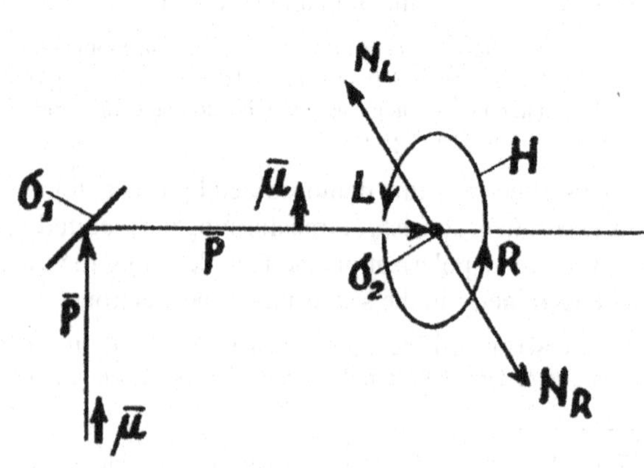

Fig. 6.9 Pictorial description of the double scattering of longitudinal electrons. *Source* De Shalit et al. (1957)

its original orientation in space. The scattered electron approaching σ_2 feels an effective magnetic field H which is caused by the current of positive nuclei of σ_2 moving towards the electron (in its rest frame). One sees easily that in the case of Fig. 2, electrons which pass to the right (R) of the nucleus add up their magnetic interaction to the electric one, whereas for those which pass to the left (L) of the nucleus the difference of the two interactions determines the net interaction. Since passing to the right implies scattering into the plane of the figure and passing to the left implies scattering out of that plane, it follows that under the conditions of Fig. 2 there will be more electrons scattered into the plane than out of it (de Shalit et al. 1957, pp. 1460–1461, emphasis in original).

In terms of the experimental apparatus, this means that "looking at σ_2 from σ_1, there will be a preferred scattering to the left for the electrons striking σ_2" (1460–1461). Which was in fact the case as determined by de Shalit and company. Still it was only realized several years later by Grodzins that this confirmation of longitudinal polarization provided the explanation and validation for Cox and Chase's discovery of an 90°/270° asymmetry (Grodzins 1959). This because their experimental apparatus and procedures could have led to the asymmetry only if the β-rays had been longitudinally polarized.

From a contemporary perspective there are, as Grodzins explains, some fine but relevant points to be made about the *optimization* of a double scatter experiment. First, that for an initially unpolarized beam of electrons what is required are very thin high Z scattering foils in order to minimize plural scattering. Second, that while this is also true in an ideal theoretical sense for an initially longitudinally polarized beam, it is virtually impossible to achieve in an actual experimental situation because of alignment problems which are exacerbated when longitudinally polarized beams are used. Thus, the de Shalit experiment used "double scattering first by thick low Z and then by a high Z scatterer" as a *compromise* that took into account the practical realities of experimental design (p. 401)." We should also note that the curved and extended shape of the first thick reflecting foil in the de Shalit experiment was designed so as to further minimize the alignment problems inherent in the use of longitudinally polarized electrons. Cox and Chase, however, used thick gold and lead foils which would not have been optimal had they used an unpolarized beam of electrons. But since the β-rays were longitudinally polarized this arrangement was near optimal at least with respect to the first scattering foil. Taking these experimental realities into account, the bottom line for Grodzins is:

It appears in retrospect that these early results [of Cox and Chase] were the first experiments showing evidence for the nonconservation of parity in weak interactions (p. 404).

Richard Cox agrees, "It appears now, in retrospect, that our experiments and those of Chase, were the first to show evidence for parity nonconservation in weak interactions (Cox 1973, p. 149)."

6.7 Summary and Conclusion

As we have seen, Cox was quick to pick up on the analogy, suggested by Darwin, between spin and the electric and magnetic fields of light, and then to devise an analog of the Barkla experiment that could be used to create and detect the polarization of electrons. All of this was accomplished with only a rudimentary understanding of the theoretical complexity and uncertainty underlying the concept of electron spin. And while unreliable Geiger counters frustrated Cox's initial experimental efforts, Chase was able to refine the apparatus and thus to achieve sufficient control over systemic uncertainty so to convincingly demonstrate the existence of polarization when comparing the 90°/270° orientations. His demonstration, however, of the 0°/180° asymmetry was admittedly less convincing.

At this stage one might have thought that once was enough to have shown that electrons could be polarized at least at the 270° orientation. But that was not to be because of Mott's contrary theoretical determination that the "greatest asymmetry" was to be found at the 0°/180° positions and that "the scattering *is* symmetrical" at the 90°/270° positions. This confrontation between experiment and theoretical prediction did, however, serve to encourage and justify further pursuit of relevant experiment and theory. But given the preponderance of the later reported null results for the 0°/180° asymmetry, such pursuit came to be restricted to the conflict of those null results with Mott's theory. In the process Chase's positive result for the 90°/270° asymmetry ceased to be of interest. Once was deemed not enough to justify pursuit of that conflict with Mott's theory.

The nearly ten-year conflict between Mott's theory and the null results for the 0°/180° asymmetry was finally resolved with Cox's discovery in 1940 of a hitherto unnoticed difference in the intensities in reflected and transmitted electron beams. And because transmission was significantly more efficient than reflection, this opened the door for the more efficient and convincing 1943 experiment by Shull, Chase and Meyers which confirmed the existence of the 0°/180° asymmetry.

But still lost in the shuffle was Chase's result for the 90°/270° asymmetry. So whatever justification there might have been in 1943 for further pursuit of electron polarization did *not* extend to solving the mystery of that asymmetry. That case remained closed. After all, on this Mott's theory had predicted a null result and its prediction of a positive result for the 0°/180° asymmetry had been confirmed by Shull, Chase and Meyers.

The discovery of parity nonconservation provided the theoretical background that led to the realization that a longitudinally polarized electron would by itself be an instance of parity violation. That was enough to justify the development of experimental methods to test whether β-rays in particular were longitudinally polarized. And so, they were, as shown by De Shalit, Lipkin and their collaborators. As a final historical irony, it was only realized several years later by Grodzins that this confirmation of longitudinal polarization provided the explanation for Cox and Chase's original 90°/270° asymmetry.

Had Cox and Chase realized in 1928 that β-rays were polarized and that transmission through the foil was significantly more efficient than reflection, and had they acted on that, then *once would have been enough both historically and justifiably so.* They did, however, come tantalizingly close. When looking for "some explanation" of the "wide divergence" among his data Cox brought up the possibility of "some asymmetry in the electron itself." And in this regard he considered:

> the supposition that the beam of β-particles *undergoes a polarization* in passing through the mica windows, similar to the polarization of light in passing through a tourmaline crystal. *This effect was in fact looked for* carefully in an experiment auxiliary to the present investigation but *it was not found* (Cox et al. 1928, p. 548, emphasis added).[12]

With regard to the increased efficiency of transmission, Chase promised after being confronted with Mott's prediction of a null result for the 90/270 asymmetry that:

> [My] next paper will deal with scattering at right angles, but the electrons will be those which have been scattered by *transmission through thin foils*, in contrast to the present case in which diffusely reflected electrons were used (Chase 1930, p. 1065, emphasis added).

But Cox's auxiliary experiment did not reveal that the β-rays were already polarized and Chase's promised transmission experiments only belatedly made an appearance in 1943. More in the category of a squandered opportunity is the fact that Chase could have run Mott's argument (for a null result at the 90°/270° orientations) backwards to conclude that β-rays were not unpolarized. But then again nobody thought to do this, not even Mott.[13] Thus while Cox and Chase got close, once wasn't really enough.

Appendix

Mott's theory of electron scattering and the associated quantum theory of spin is not something that can be profitably reviewed in a brief appendix. What we can do is

---

[12]Unfortunately, Cox did not provide an account of the nature of this "auxiliary" experiment. Given the methodology for dealing with systemic uncertainty he used elsewhere we suspect that the mica windows used to ferret out slow moving electrons were varied and the resultant perturbation of the resulting asymmetries at the 90°/270° orientations did not suggest that anything so major as an initial polarization was involved.

[13]For some very provocative reflections on this lost opportunity written from the perspective of "a not very abstract-minded experimental physicist, or (shall I say) as it might have appeared to Rutherford had he lived to hear the end of the story" see (Blackett 1959). Writing from this perspective Patrick Blackett notes that "it seems most likely that the [Cox and Chase] experiments really did demonstrate the *longitudinal polarization* of natural β-particles and so, in modern jargon, demonstrated the non-conservation of parity. *If these experiments had been followed up*, for instance, by any of us younger colleagues of Rutherford, many of whom were at that time looking for new and simple experiments to do. .. then the hypothesis of the universal conservation of parity would never have been put forward and the essential asymmetry of nature between right- and left-handed systems would have been established nearly thirty years earlier than it in fact was" (p. 510, emphasis added).

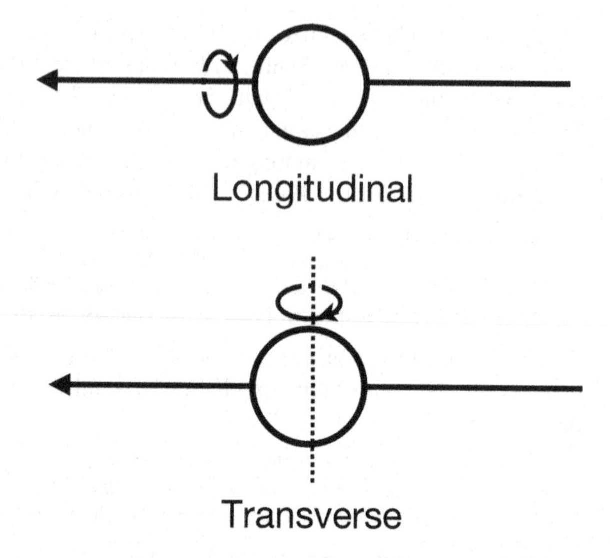

Fig. 6.10 Longitudinal and transverse orientation of electron spin axis. *Source* Authors

to present the rudiments of an elementary spinning magnet analogy for spin. While inadequate when it comes to the quantitative details, it does work reasonably well as a *qualitative* account that is helpful in understanding the electron double-scatter experiments discussed in this chapter.

The basic idea is to conceive the electron as a small spinning magnet. Classically understood, magnets produce a magnetic field and if placed in a powerful magnetic field will orient in a way consistent with the external field. But unlike ordinary spinning magnets, electron spin does not come in continuous amounts. There is only spin up and spin down. And in the case of electrons the only values are ½ and minus ½, where the former refers to clockwise rotation and the latter to counter-clockwise rotation with respect to the velocity of the electron. There are in addition two orientations for the spin axis: (1) the axis of the spin is perpendicular to the velocity, and (2) the axis is parallel to the velocity. See Fig. 6.10.

An electron beam is said to be *polarized* (with respect to a given reference direction) when the two spin states of the constituent electrons do not occur in equal amounts. The *degree* of polarization (*P*) is given by the fractional difference of the two spin state populations n_+ and n_-:

$$p = \frac{n_+ - n_-}{n_+ + n_-}$$

If the direction of the spin is perpendicular to the momentum of the electron beam, this equation defines the *transverse polarization* of the beam. If, on the other hand, the direction of the spin is parallel to the momentum, this equation defines the *longitudinal polarization* of the beam.

What happens then when an electron comes into contact with the foils of a double scatter experiment? If the electron were truly a spinning magnet its orientation would

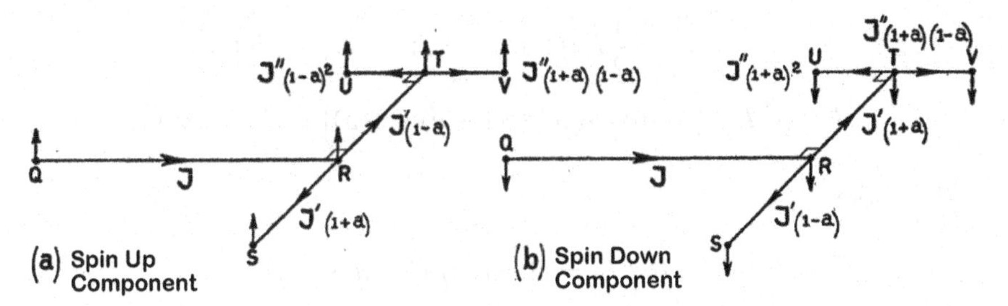

Fig. 6.11 The intensities are given for double scattering (at right angles) and spin orientations "up" and "down". The beams are scattered at R and T. Intensities for double scattering of an unpolarized beam are obtained by taking the sum so that different intensities result in U and V. *Source* Toelhoek (1956)

be changed by the interaction with the more powerful magnetic field. The analogous process for real electrons results in a change of the spin distribution for the electron beam. The resultant force of the interaction between the electron spin and the magnetic field is known as the *spin-orbit coupling value* and depends on the relative orientation of the electron spin and the magnetic field.

For the time being, we'll restrict attention to where the axis of our small magnet's spin is perpendicular to its velocity and where the result of an interaction with a scatter foil is a change in the transverse polarization of the electron beam.[14] See Fig. 6.11 where an initially unpolarized electron beam undergoes transformations in spin polarization with respect to various configurations of a double scatter apparatus. The left side (a) displays these changes *with respect to the spin up distribution*, while the right side (b) shows the changes *with respect to the spin down distributions.*

The path defined by *QRTU* represents that of the Cox-Chase apparatus at the 180° orientation with the β-ray source at Q and the electron counter at U. The path defined by *QRTV* represents that of the Cox-Chase apparatus at the 0° orientation with the counter at V. For current purposes the expressions $\mathbf{J}^{\mathbf{i}}(1 \pm a)$ may be understood as representing the incremental change in spin polarization at the various segments of the electron path, where these changes are a function of $(1 \pm a)$ and where the value of a depends on the experimental particulars. Alternatively, $\mathbf{J}^{\mathbf{i}}(1 \pm a)$ may be understood as representing the signal intensities at the various segments of the electron path.[15] Thus, for example, the expression $\mathbf{J'}(1 - a)$ on the path from R to T on the left side of the figure registers the change in spin up polarization (and the corresponding change in intensity) at that stage of the process.

Now *superimpose* the spin up and spin down diagrams to obtain the composite relative polarization differences at U and V which translate into measurable intensity differences as follows (where I" is a constant):

---

[14]We're also implicitly assuming that the velocity vectors of the electrons in the initial beam are uniformly aligned and are perpendicular to the axis of the first scatter foil. Variation from this idealized assumption is dealt with in more developed theoretical accounts.

[15]For details and the experimental basis for the determination of these values see (Toelhoek 1956, 282–287) and for a similar but less developed version see (Grodzins 1959, 399–401).

$$I_U = \frac{1}{2} I''\left[(1+a)^2 + (1-a)^2\right] = I''(1+a^2)$$

$$I_V = \frac{1}{2} I''[(1+a)(1-a) + (1+a)(1-a)] = I''(1-a^2)$$

Therefore,

$$I_U / I_V = (1+a^2)/(1-a^2)$$

Thus, for an initially unpolarized (transverse) beam, there will be an asymmetry where the intensity will be larger at U (the 180° orientation) than at V (the 0° orientation)—which was experimentally confirmed in Shull et al. (1943) as discussed in Sect. 6.5.

In order to see what happens for the 90° and 270° orientations simply rotate the U-V axis 90° vertically around T. Because the spin polarization (and intensity) differences occur *only* along the horizontal unrotated U-V axis, this means that there will be no polarization difference along the now vertical U-V axis.

In sum, we now have a least a qualitative explanation of the Mott results, namely, that "the greatest symmetry" will occur between the 0° and 180° orientations, and that "the scattering *is* symmetrical" between the 90° and 270° orientations. As the reader may recall, Mott also concluded that "the [90°/270°] asymmetry found by [Cox and Chase] must be due to some other cause (Mott 1929, p. 431)." While Mott was right about there being "some other cause," he didn't realize that this other cause was that the initial electron beam was longitudinally polarized and not, as assumed by Cox and Chase, an unpolarized transverse beam.

We'll now take a quick look at how using a longitudinally polarized beam affects the analysis of any resultant asymmetries. Because β-rays are longitudinally polarized this means that when they make contact with the first scatter foil their velocity will rotate 90° but *the spin axis will not follow suite.* See (Toelhoek 1956, p. 283). As a consequence the axis of spin rotation will shift from longitudinal to transverse (i.e., from being parallel to momentum to being perpendicular). See Fig. 6.12. This significantly changes the analysis of the expected asymmetry in a double scattering experiment. Without going into the technical details the general result is that the there will be asymmetries for *both* the 0°/180° and the 90°/270° comparisons and that the asymmetry will be larger for the 0°/180° comparison than for the 90°/270° comparison. This is somewhat at odds with Chase's result which had it the other way round, though here it must be noted that any specific determination of the expected asymmetries will be sensitive to the particulars of the experiment including among other things the electron velocities that were achieved.[16]

As we note in Sect. 6.8, until the discovery of parity non-conservation no one thought of running the Cox and Chase experiment "backwards" to conclude that the initial beam was not unpolarized. This would have been just a straightforward contrapositive of the Cox and Chase experimental results. Once though there was

[16]For further discussion of the conflict and Cox's ambivalent reaction see (Franklin 1979, 228–229).

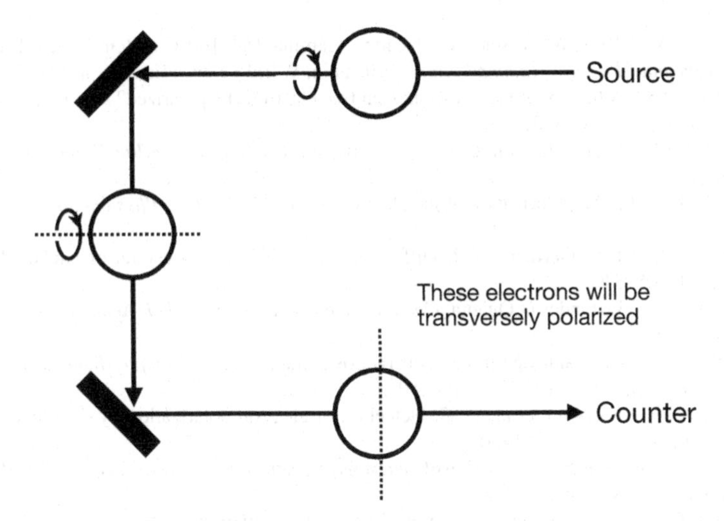

Fig. 6.12 Conversion at first scatter foil of longitudinally polarized electron to transverse orientation. *Source* Authors

a specific theoretical determination of the asymmetries that would result when the electron beam was initially longitudinally polarized then one could argue (given the absence of an alternative account) that the initially assumed value or range of values for the initial polarization was likely to be correct. Thus, as concluded in De Shalit et al. (1957) "It can, however, be seen easily that our results are *compatible* with 100% polarization of the initial beam. The direction of polarization is unambiguous (emphasis added),"[17]. Considerable effort has since been expended to further refine this sort of backward determination of the polarization of β-rays but that for us is a subject for another time.

References

Barkla, C.G. 1906. Polarization in secondary Röntgen radiation. *Proceedings of the Royal Society of London* A77 (516): 247–255.

Blackett, P.M.S. 1959. Non-conservation of parity. *American Scientist* 47 (4): 509–514.

Chase, C.T. 1930a. The scattering of fast electrons by metals I. *Physical Review* 36: 984–987.

Chase, C.T. 1930b. The scattering of fast electrons by metals II. *Physical Review* 36: 1060–1065.

Chase, C.T., and R.T. Cox. 1940. The scattering of 50-kilovolt electrons by aluminum. *Physical Review* 58: 243–251.

Cox, R.T., C.G. McIlwraith, et al. 1928. Apparent evidence of polarization in a beam of β-rays. *Proceedings of the National Academy of Sciences (USA)* 14: 544–549.

[17]There is a technical nicety that should be noted, namely, that in this context "full [i.e. 100%] polarization denotes a longitudinal polarization of v/c" (Lipkin et al. 1958, 224 n.10). Here v is the electron velocity and c the speed of light, and where this ratio is a theoretically derived limitation on the amount of polarization that's possible.

Cox, R.T.B.B.M.P. 1973. World Science Education. Gamma, 154–160. Cox Reminisces. *Adventures in Experimental Physics, Gamma Volume*. Princeton. World Science Education, 149.
Darrigol, O. 1984. A history of the question: Can free electrons be polarized? *Historical Studies in the Physical Sciences* 15: 39–79.
Darwin, C.G. 1927. The electron as a vector wave. *Proceedings of the Royal Society (London)* A116: 227–253.
Darwin, C.G. 1930. The polarization of the electron. *Proceedings of the Physical Society* 42: 379–384.
Davisson, C., and L.H. Germer. 1927. Diffraction of electrons by a crystal of Nickel. *Physical Review* 30: 705–740.
De Broglie, L. 1923a. Ondes et Quanta. *Comptes Rendus des Seances de L'Academie des Sciences* 177: 507–510.
De Broglie, L. 1923b. Quanta de Lumiere, diffraction et interferences. *Comptes Rendus des Seances de L'Academie des Sciences* 177: 548–550.
De Shalit, A., S. Cuperman, et al. 1957. Detection of electron polarization by double scattering. *Physical Review* 107: 1459–1460.
Dymond, E.G. 1932. On the polarisation of electrons by scattering. *Proceedings of the Royal Society (London)* A136: 638–651.
Dymond, E.G. 1934. On the polarization of electrons by scattering. II. *Proceedings of the Royal Society (London)* A145: 657–668.
Franklin, A. 1979. The discovery and nondiscovery of parity nonconservation. *Studies in History and Philosophy of Science* 10: 201–257.
Frauenfelder, H., R. Bobone, et al. 1957. Parity and the polarization of electrons from Co^{60}. *Physical Review* 106: 386–387.
Goertzel, G., and R.T. Cox. 1943. The effect of oblique incidence on the conditions for single scattering of electrons by thin foils. *Physical Review* 63: 37–40.
Grodzins, L. 1959. The history of double scattering of electrons and evidence for the polarization of beta rays. *Proceedings of the National Academy of Sciences (USA)* 45: 399–405.
Kikuchi, K. 1940. On the polarization of electrons. *Proceedings of the Physico-Mathematical Society of Japan* 22: 805–824.
Langstroth, G.O. 1932. Electron polarisation. *Proceedings of the Royal Society (London)* A136: 558–568.
Lee, T.D., and C.N. Yang. 1957. Parity nonconservation and a two-component theory of the neutrino. *Physical Review* 105: 1671–1675.
Lipkin, H.J., S. Cuperman, et al. 1958. Measurement of beta-ray polarization of Au^{198} by double coulomb scattering. *Physical Review* 109: 223–224.
Mott, N.F. 1929. Scattering of fast electrons by atomic nuclei. *Proceedings of the Royal Society (London)* A124: 425–442.
Mott, N.F. 1932. The polarisation of electrons by double scattering. *Proceedings of the Royal Society (London)* A135: 429–458.
Richter, H. 1937. Zweimaliger Streuun schneller Elektronen. *Annalen der Physik* 28: 533–554.
Rupp, E. 1929. Versuche zur Frage nach einer Polarisation der Elektronenwelle. *Zeitschrift fur Physik* 53: 548–552.
Rupp, E. 1930. Ueber eine unsymmetrische Winkelverteilung zweifach reflektierter Elektronen. *Zeitschrift fur Physik* 61: 158–169.
Rupp, E. 1931. Direkte Photographie der Ionisierung in Isolierstoffen. *Naturwissenschaften* 19: 109.
Rupp, E. 1932a. Versuche zum Nachweis einer Polarisation der Elektronen. *Physikalische Zeitschrift* 33: 158–164.
Rupp, E. 1932b. Neure Versuche zur Polarisation der Ele4ktronen. *Physikalische Zeitschrift* 33: 937–940.
Rupp, E. 1932c. Ueber die Polarisation der Elektronen bei zweimaliger 90°-Streuung. *Zeitschrift fur Physik* 79: 642–654.

Rupp, E. 1934. Polarisation der Elektronen an freien Atomen. *Zeitschrift fur Physik* 88: 242–246.
Rupp, E. 1935. Mitteilung. *Zeitschrift fur Physik* 95: 801.
Rupp, E., and L. Szilard. 1931. Beeinflussung 'polarisierter' Elektronenstrahlen durch Magnetfelder. *Naturwissenschaften* 19: 422–423.
Shull, C.G. 1942. Electron polarization. *Physical Review* 61: 198.
Shull, C.G., C.T. Chase, et al. 1943. Electron polarization. *Physical Review* 63: 29–37.
Thomson, G.P. 1934. Experiment on the polarization of electrons. *Philosophical Magazine* 17: 1058–1071.
Toelhoek, H.A. 1956. Electron polarization, theory and experinebt. *Reviews of Modern Physics* 28 (3): 277–298.
Uhlenbeck, G.E., and S.A. Goudsmit. 1926. Spinning electrons and the structure of spectra. *Nature* 117: 264–265.

Chapter 7
The Discovery of Parity Nonconservation

In Chap. 6 we discussed the experiments of Richard Cox and his collaborators and that of his student Carl Chase. These experiments, at least in retrospect, showed that electrons from beta decay were longitudinally polarized, and thereby that parity was not conserved. No one in the physics community realized this at the time. Although the results were seen as anomalous no replications were attempted. The reasons for that neglect included a lack of the necessary theoretical context and the fact that other experiments using the same type of apparatus were deemed more important.

The situation was dramatically different in the 1950s. The physics community was faced with a problem, the $\theta - \tau$ puzzle, which resisted solution using accepted theory. In 1956, Tsung-Dao Lee and Chen-Ning Yang proposed that the $\theta - \tau$ puzzle would be solved if parity was not conserved in the weak interactions, a startling proposal. They also suggested several experiments that would test that hypothesis. In early 1957 three different experimental results were reported at approximately the same time which collectively and decisively demonstrated that parity was not conserved.[1] In this chapter we will discuss in detail only one of the experiments, that performed by Chien-Shiung Wu and her collaborators on the beta decay of oriented nuclei, and only briefly discuss the other two. We chose this experiment because it was the most direct and consequently required the least complicated and theoretically dependent apparatus.

7.1 Background

The parity transformation (aka operation) is defined as the reflection of all spatial coordinates through the origin (i.e. $x \rightarrow -x$, $y \rightarrow -y$, $z \rightarrow -z$; i.e. x is replaced by –

[1]The papers by Wu and collaborators (1957) and that of Garwin et al. (1957) were published in the February 15, 1957 issue of *Physical Review* whereas that of Friedman and Telegdi (1957) did not appear until the March 1 issue.

A. Franklin and R. Laymon, *Once Can Be Enough*,
https://doi.org/10.1007/978-3-030-62565-8_7

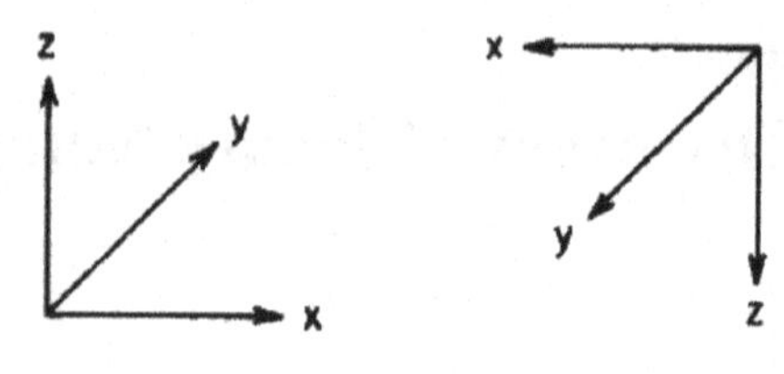

Fig. 7.1 Right-handed and left-handed coordinate systems. *Source* Authors

x, y by −y, and z by −z). It is also the difference between left- and right-handed coordinate systems (Fig. 7.1). An observer transported from the original reference frame to the parity mirror frame and subjected to the corresponding rotations would "see" the left-right, up-down and front-back orientations reversed but with the original *relative* orientations (among the items observed) intact. Such *space inversion invariance* has been held to be a requirement for the laws of physics. Thus, for example, the parity mirror images of a prediction and the corresponding experimental result must give the same result as the original prediction and experiment.

All students of introductory physics are taught that if you want to find the force between two parallel currents you first find the magnetic field due to one of the currents using a right-hand rule, and then calculate the force exerted by that field on the second current using a second right-hand rule. If you are interested only in the observables, current and force, then two left-hand rules will work equally well. This is parity conservation in classical electromagnetic theory.

In quantum mechanics we speak of the behavior of the wave functions describing particles under the parity operation. We find that $\psi(\mathbf{r}) = \pm\psi(-\mathbf{r})$, where $\mathbf{r}$ is the vector describing a position in three-dimensional space and where "$-\mathbf{r}$" denotes that the parity transformation has been applied.[2] If $\psi(\mathbf{r}) = -\psi(-\mathbf{r})$ we speak of the wave function as having odd or negative parity. If $\psi(\mathbf{r}) = \psi(-\mathbf{r})$ the parity is even or positive. These transformations leave $|\psi(\mathrm{r})|^2$, the probability of finding the particle in a region, unchanged.[3] Thus even though space inversion invariance is not satisfied for odd wave functions, that difference between even and odd functions does not manifest itself when it comes to what's observable because what's observable (spatial distribution of mass and charge) depends on $|\psi(\mathrm{r})|^2$. And because of this what's observable satisfies the requirement of *space inversion invariance* because $|\psi(\mathrm{r})|^2$ has the same value in both the original and the *parity transformed* mirror image. If a system contains more than one type of particle then one can observe the relative phase.[4]

[2]The notation "$-\mathbf{r}$" reflects the fact that by and large polar coordinates are used in physics and that reversing the direction of the $\mathbf{r}$ vector automatically induces the parity transformation (expressed in polar coordinates) because there's a corresponding reversal in the angular coordinates.

[3]The use of the absolute value notation (|…|) is meant to succinctly express the following: If $\psi(\mathbf{r})$ is odd then $(\psi(\mathbf{r}))^2 = (-\psi(-\mathbf{r}))^2 = (\psi(-\mathbf{r}))^2$. On the other hand, if $\psi(\mathbf{r})$ is even then $(\psi(\mathbf{r}))^2 = (\psi(-\mathbf{r}))^2$. Thus, in either case, $(\psi(\mathbf{r}))^2 = (\psi(-\mathbf{r}))^2$.

[4]If a particle has angular momentum l then the parity of the wave function is $(-1)^l$.

The modern history of parity conservation began with the work of Laporte (1924). In studying the spectrum of radiation emitted by iron atoms, Laporte found that he could classify the states of iron atoms into two classes and that radiation was emitted only when an atom made a transition from a state in one class to a state in the other class. Intraclass transitions did not occur.

Wigner (1927) provided an explanation of this rule based on the assumption of *parity conservation* in the sense that the intrinsic parity of the initial state must be equal to the product of the intrinsic parities of the components of the final state of the process. Wigner explained that the two classes of states identified by Laporte were states of positive and negative parity. Because the intrinsic parity of the photon, in the most important case of electric dipole radiation, is negative, then in order for the total parity of the system to be conserved, the parity of the atomic state must change. Thus, we get radiation emitted only when the atom makes a transition from a state of positive parity to one of negative parity, or vice versa. Transitions between states of the same parity of are forbidden by parity conservation. Parity conservation was quickly accepted as an established principle of physics. As Frauenfelder and Henley later remarked, "Since invariance under space reflection is intuitively so appealing (why should a left- and right-handed system be different?), conservation of parity quickly became a sacred cow (Frauenfelder and Henley 1975, p. 359)."[5]

In the early 1950s several physicists began speculating about the possibility of the violation of symmetry principles. The first of these was in a paper dealing with the possibility of electric dipole moments for elementary particles and nuclei by Purcell and Ramsey. They noted that one of the major theoretical arguments against the existence of such electric dipole moments rested on the conservation of parity which had not yet been tested for elementary particles and nuclei.

> The argument against electric dipoles, in another form, raises directly the question of parity. A nucleon with an electric dipole moment would show an asymmetry between left- and right-handed coordinate systems; in one system the dipole moment would be parallel to the angular momentum and in the other, anti-parallel.
> But there is no compelling reason for excluding this possibility. It would not be the only asymmetry of particles of ordinary experiences, which already exhibit conspicuous asymmetry in respect to electric charge (Ramsey 1956, p. 807)."[6]

There is a further suggestion of the possibility of parity nonconservation in a paper by Wick et al. (1952). They discussed the concept of intrinsic parity of elementary particles and point out that there are limitations to this concept.

---

[5]An amusing early reference to parity conservation occurred in a paper by Jordan and Kronig (1927). In this paper Jordan and Kronig noted that the chewing motion of cows is not straight up and down, but is rather either a left-circular or a right-circular motion. They reported on a survey of cows in Sjaelland. Denmark, and observed that 55% are right-circular and 45% left-circular, a ratio they regard as consistent with unity.

[6]Lee and Yang later cited this paper, but as Lee later told Franklin they used a secondary source and had not read the original paper.

> The purpose of this paper is to point out the possible (and in certain cases necessary) existence of limitations to one of these general concepts, the concept of "intrinsic parity" of an elementary particle. Even though no radical modification of our thinking is thereby achieved, we believe that the injection of a certain amount of caution in this matter may be useful, as it may prevent one from calling "theorems" certain assumptions or from discarding as "impossible" forms of the theory, which under a more flexible scheme are perfectly consistent (p. 101)

By mentioning these earlier suggestions we are not in any way implying that they influenced Lee and Yang or that they are precursors in any real sense. They show only that by 1950 there was at least a consideration of the possibility of such symmetry violations. It is still a crucial and major step to not only suggest that such violations will solve a physical problem, namely the $\theta - \tau$ puzzle, but to discuss the experimental tests of such a suggestion, as Lee and Yang did.

Other important developments came from the study of V particles discovered in 1947 by Rochester and Butler (1947). These were neutral particles which decayed into two charged particles and appeared as V's in cloud chamber photographs (Fig. 7.2). Further work showed that there were two types of V particles; (1) lower mass particles, approximately one half the mass of the proton, later called K mesons, and (2) particles with mass greater than that of the protons called hyperons.

From their discovery in 1947, by Rochester and Butler, the neutral K mesons, or kaons, have exhibited odd features. One puzzling observation about the kaons was put this way by C. N. Yang: "while the[se] strange particles are produced quite abundantly (say 5% of the pions) at GeV energies and up, their decays into pions and nucleons are rather slow (10^{-10} s). Since the time scale of pion-nucleon interactions is of the order of 10^{-23} s, it was very puzzling how to reconcile the abundance of these objects with their longevity (10^{13} units of time scale) (Yang 1956)." In 1952 Abraham Pais suggested that strange particles appear in the strong interactions only in pairs, which he called "associated production." (Pais 1951). When they decay, however, they necessarily do so one at a time, apparently through weak interactions, which occur more slowly than strong ones. In an ad hoc way, at least, this explained the phenomena. Pais' suggestion was elaborated in the theory of strangeness conservation put forward by Gell Mann and Nishijima between 1953 and 1955 (Nakano and Nishijima 1953; Nishijima 1954, 1955; Gell-Mann 1953, 1956). They supposed that K mesons and hyperons (particles heavier than nucleons) possessed a quantity called strangeness, which is conserved in strong and electromagnetic interactions, but not in weak interactions,[7] and so accounts for the copious production and long decay time of "strange" particles. The conservation of strangeness forbids the occurrence of certain strong reactions and decays otherwise allowed. No violation of this restriction had been noticed by 1956, or at any time since.

One further problem with these strange particles what was known as the "$\theta - \tau$" puzzle. There were two elementary particles, the τ and the θ (they were K mesons). On one set of criteria, namely mass and lifetime, they appeared to be the same particle.

---

[7]This was an example of a conservation law that applied in some forms of interaction, but not in others. Franklin asked Lee if the idea of such a partial conservation law had any influence on his suggestion of parity nonconservation. Lee replied that it didn't (private conversation with Franklin, 1978).

Fig. 7.2 An example of the decay of a V-particle. The decay is seen near the top of the lower chamber. *Source* Alford and Leighton (1953)

On another set of criteria, spin and parity, they seemed to be different particles. Physicists, at the time, attempted to solve the $\theta - \tau$ puzzle within the framework of accepted theories. One of the earliest was by Lee and Orear, who in 1955, suggested that perhaps the heavier of the two particles might decay into the lighter one and thereby give results consistent with experiment.

> From the analysis of Dalitz it is becoming necessary to assume that the $K^{\pm}_{\pi 3}$ $(\tau^{\pm})$ and $K^{\pm}_{\pi 2}$ $(\theta^{\pm})$ are due to at least two different charged particles. Recent experimental evidence encourages the hypothesis that the observable lifetimes are exactly the same (at least after travel times > 10^{-9} s). As a solution to this possible dilemma we propose that there exist two heavy mesons, the $\theta^{\pm}$ and the $(\tau^{\pm})$, where $(\theta^{\pm} \rightarrow \pi^{\pm} + \pi^{o}$ and $\tau^{\pm} \rightarrow \pi^{\pm} + 2\pi$. We propose further that the heavier of these two has a lifetime of 10^{-8} s with a significant branching ratio for gamma decay to the lighter one. Finally, if we assign a lifetime of order 10^{-9} s or less to the lighter one, we achieve

a situation where both particles have different lifetimes, but in most experiments appear to have exactly the same lifetime (Lee and Orear 1955).

At the 1956 Rochester Conference, Alvarez reported on a search for the γ-rays which would arise from the proposed Lee and Orear decay schemes, $\theta^{\pm} \rightarrow \tau^{\pm} + 2\gamma$ and vice versa, or $\tau^{\pm} \rightarrow \theta^{\pm} + \gamma$, and concluded, "Therefore, no Lee-Orear γ-rays were seen with pulses of the order of 0.5 MeV (in 1956, V. 28–30).

At the same conference Marshak suggested that a last effort be made to explain the θ and τ as a single particle using larger spin values. He reported that the lowest spin value compatible with the single particle assumption was 2. Although this was compatible with the data existent at the time, Orear and collaborators subsequently showed that this was extremely unlikely (Orear et al. 1956). There were numerous other attempts at a solution. All of them were unsuccessful.

Lee and Yang realized that the puzzle could be solved if parity were not conserved in the weak interaction, the interaction responsible for radioactive decay and for the decay of many elementary particles. "It finally became clear to me, in early 1956, that the solution to the $\theta - \tau$ puzzle must lie in something much deeper: perhaps, parity is not conserved, and the θ and τ are indeed the same particle (Lee 1971, p. 10)." Lee and Yang examined the existing evidence for parity conservation and found, to their surprise, that although there was good evidence for parity conservation in the strong and electromagnetic interactions, there was, in fact, no evidence available for the weak interactions. Lee reported their search as follows, "I then borrowed from Wu the authoritative book on β-decay edited by K. Siegbahn, and proceeded with Yang to calculate systematically all parity violating effects.... After we went through the entire Siegbahn book, rederived all these old formulas with this new interaction, it became obvious to us that not only was there, at the time, not a single evidence for parity conservation in β-decay but that we must have been very stupid?... Once we stopped calculating and started to think, in a rather short time, it dawned on us that the reason for this lack of evidence was the simple fact that nobody had made any attempt to observe a physical pseudoscalar[8] from an otherwise seemingly right-left symmetric arrangement (Lee 1971, p. 22)."[9]

7.2 The Discovery

A. The Proposal

Lee and Yang began their paper with a statement of the problem.

---

[8]A pseudoscalar is a quantity that changes sign under a parity operation

[9]It is not surprising that Lee and Yang missed the experiments of Cox et al. and of Chase. They were looking at experiments involving beta decay. Those experiments were on electron polarization and would not have been included in Siegbahn's book.

Recent experimental data indicate closely identical masses and lifetimes of the θ^+ ($\equiv K^+_{\pi 2}$) and the τ^+ ($\equiv K^+_{\pi 3}$) mesons.[10] On the other hand, analyses of the decay products of the τ^+ strongly suggest on the grounds of angular momentum and parity conservation that the τ^+ and the θ^+ are not the same particle. *This poses a rather puzzling situation that has been extensively discussed* (Lee and Yang 1956, p. 254)."

They then offered their radical solution to the problem.

One way out of the difficulty is to assume that parity is not strictly conserved, so that θ^+ and τ^+ are two different decay modes of the same particle, which necessarily has a single mass value and a single lifetime. We wish to analyze this possibility in the present paper against the background of the existing experimental evidence of parity conservation. It will become clear that existing experiments do indicate parity conservation in strong and electromagnetic interactions to a high degree of accuracy, but that for the weak interactions (i.e., decay interactions for the mesons and hyperons, and various Fermi interactions) parity conservation is so far only an extrapolated hypothesis unsupported by experimental evidence (Lee and Yang 1956, p. 254).

The suggestion of parity nonconservation solved the "$\theta - \tau$" puzzle. One assumes parity conservation and that the θ and τ are the same particle because of they have the same mass and lifetime. One then determines the parity by looking at the final state after decay then one gets a contradiction. The θ and τ have spin zero so by conservation of angular momentum the spin of the final state must be zero. The parity of the final state is thus determined by the intrinsic parity of the particles in the final state. The parity of the final state for the θ, which decays into two pions, is $(-1)^2 = +1$, or positive. For the τ, which decays into three pions, the parity of the final state is $(-1)^3 = -1$, or negative. If parity isn't conserved then the θ^+ and τ^+ are merely two different decay modes of the same particle.

Lee and Yang went on to state,

One might even say that the present θ - τ puzzle may be taken as an indication that parity conservation is violated in weak interactions. This argument is, however, not to be taken seriously because of the paucity of our knowledge concerning the nature of strange particles. It supplies rather an incentive for an examination of the question of parity conservation. To decide unequivocally whether parity is conserved in weak interactions, one, must perform an experiment to determine whether weak interactions differentiate the right from the left (Lee and Yang 1956, p. 254).

The proposal by Lee and Yang was met with some skepticism by the physics community. Wolfgang Pauli, a Nobel Prize winner wrote to Victor Weisskopf, "I do *not* believe that the Lord is a weak left-hander, and I am ready to bet a very large sum that the experiments will give symmetric results (quoted in Bernstein 1967, p. 59). In a letter to C.S. Wu on January 19, 1957, after hearing word of the results of her experiment he noted, "I did not believe in it [parity nonconservation] when I read the paper of Lee and Yang (Quoted in Maglic 1973, p. 122)." Lee reported that Felix Bloch, another Nobel Prize winner, offered to bet other members of the Stanford Physics Department his hat, that parity was conserved (private communication to Franklin, 1978.

---

[10]The θ^+ decays into two pions ($\theta^+ \rightarrow \pi^+\pi^o$ and the τ^+ into three pions ($\tau^+ \rightarrow \pi^+\pi^o\pi^o$ or $\tau^+ \rightarrow \pi^+\pi^+\pi^-$.

B. The Experiments

1. The Beta Decay of Oriented Nuclei

In sum, having proposed that "parity conservation is violated in weak interactions," the problem was to devise some experiments that would "determine whether weak interactions differentiate the right from the left" (Lee and Yang 1956, p. 254). And in fact, Lee and Yang came up with several such proposals. The simplest and most direct was to determine whether β decay from an oriented, i.e., polarized nucleus, was distributed in a way that did not satisfy the parity transformation requirement.

> A relatively simple possibility is to measure the angular distribution of the electrons from β decays of oriented nuclei. If θ is the angle between the orientation of the parent nucleus and the momentum of the electron, an asymmetry of distribution between θ and $180° - \theta$ constitutes an unequivocal proof that parity is not conserved in β decay. If the angular distribution were symmetrical that would argue for parity conservation. . . . To be more specific, let us consider the allowed β transition of any oriented nucleus, say Co^{60}. The angular distribution of the β radiation is of the form
>
> $$I(\theta)d\theta = (\text{constant})\,(1+\alpha\cos\theta))\sin\theta d\theta$$
>
> where α is proportional to the interference term . . . . If $\alpha \neq 0$, one would then have a positive proof of parity nonconservation in β decay. . . . It is noteworthy that in this case the presence of the magnetic field used for orienting the nuclei would *automatically cause a spatial separation* between the electrons emitted with $0 < 90°$ and those with $0 > 90°$. Thus, this experiment may prove to be quite feasible, emphasis added).

The proposal raises two questions. First, the non-expert reader may wonder how the proposal test "constitutes an unequivocal proof that parity is not conserved in β decay." And second, exactly how was this "relatively simple possibility" to be realized in a real-world experiment in such way as to convincingly decide the issue? Relatively simple in concept does not guarantee relatively simple to experimentally implement.

With regard to the first question, we'll need to introduce some additional background theory in order to specifically connect the experimental result with parity nonconservation. It can be shown that:

> If a decay process is not invariant under parity space inversion then parity conservation does not hold.

While the proof is much too complex to be dealt with here, we can give some idea as to the basic strategy involved.[11] It will be helpful to introduce the following notation. Let $W(\boldsymbol{p}, \boldsymbol{S})$ stand for the probability of a transition from the initial state to the final state of a decay process where that probability will depend on all the momenta $\boldsymbol{p}$ (initial and final) and all the spins $\boldsymbol{S}$. What can be shown then is that if parity is conserved, and given the symmetry properties of parity (construed as an operator on the wave equation) then $W(\boldsymbol{p}, \boldsymbol{S}) = W(-\boldsymbol{p}, \boldsymbol{S})$ where $-\boldsymbol{p}$ is the parity transformation of the momenta.[12] In other words, parity conservation requires the

---

[11] For a complete and precise rendition see Gibson and Pollard (1976, pp. 119–127, 160–162)

[12] The parity transformation does not change the spin, i.e., $-\boldsymbol{S} = \boldsymbol{S}$.

equivalence of the transition probabilities in the real and the parity mirror spaces. More specifically, and relevant for the experiment proposed by Lee and Yang, in order to test for the equivalence of $W(\boldsymbol{p}, \boldsymbol{S})$ and $W(-\boldsymbol{p}, \boldsymbol{S})$ it will be sufficient to test whether the pseudo-scalar $(\boldsymbol{S}^*\boldsymbol{p})$ changes value once transformed into the parity mirror world.[13] If its value changes then the probabilities will differ insofar as there will be a different configuration of spin and momenta. Thus parity conservation will not hold because $W(\boldsymbol{p}, \boldsymbol{S}) \neq W(-\boldsymbol{p}, \boldsymbol{S})$. As we'll see in the case of Co^{60} and beta decay the relevant difference was the change in the dot product of the spin of the Co^{60} nucleus and the asymmetric distribution of the beta particle momenta.

We'll need two last bits of background theory. The simple transformation of one coordinate to its negative is called a "reflection" or "simple mirror" transformation, e.g., (x to $-x$, y to y, z to z). The relevant fact here is that if a reflection transformation yields a non-congruent image then the parity transformation (x to $-x$, y to $-y$, z to $-z$) will also yield a non-congruent image. Thus, if $(\boldsymbol{S}^*\boldsymbol{p})$ changes value in a reflection transformation, then it will also change value in the parity mirror world. We also need to emphasize that spin is what's known as an "axial vector" which means that while the orientation of its rotation changes under a reflection transformation (e.g., from clockwise to counterclockwise), the parity transformation (as can be readily seen) does not alter the rotation orientation (e.g. clockwise stays clockwise).

With this necessary background in place, we can now turn to the question of how the "relatively simple possibility" proposed by Lee and Yang was realized in a real-world experiment. Enter Chien-Shiung Wu, Ernest Ambler, Raymond Hayward, Dale Hoppes, and Ralph Hudson. First on their agenda was how to orient, i.e., polarize, the Co^{60} nuclei in a way that could be both *monitored* and *isolated* from systemic interference. As we shall see, a determinate and stable polarization state was necessary in order to show that the parity mirror state was not congruent with the real-world state. In this regard it was known that "Co^{60} nuclei can be polarized by the Rose-Gorter method in cerium magnesium (cobalt) nitrate, and the degree of polarization detected by measuring the anisotropy of the succeeding gamma rays (Wu et al. 1957, p. 1413)." In short, the method of polarization was to prepare certain structurally advantageous salts containing Co^{60} that under significant cooling could be polarized by the imposition of a magnetic field.

That the Co^{60} had been successfully polarized could be determined by the corresponding anisotropy of the ejected γ-radiation. For the details see Ambler et al. (1953). Thus, "[t]he observed gamma-ray anisotropy was used as a measure of polarization, and, effectively, temperature" (Wu et al. 1957, p. 1413). This correspondence between the γ-radiation anisotropy and the Co^{60} polarization played a crucial role in conducting the experiment because a continuing anisotropy provided a reliable indication that the Co^{60} remained polarized during the course of the experiment.

In addition to the general problem of achieving and maintaining the very low temperatures required, there were "two [related] major difficulties" with regard to the apparatus that had to be overcome: "The beta-particle counter should be placed

[13] For reader who may have forgotten the dot product of two vectors is the product of the length of the vectors and the cosine of the angle between the vectors.

inside the demagnetization cryostat, and the radioactive nuclei must be located in a thin surface layer and polarized" (Wu et al. 1957, p. 1413). The critical arrangement of these elements in the "cryostat" apparatus is depicted in Fig. 7.3. As can be seen, the Co^{60} is located at the lower end of the cryostat while a thin anthracene crystal, located above the Co^{60}, was used to detect the beta particles where the scintillations were transmitted to a photomultiplier. In addition, gamma ray detectors (denoted "NaI") were located as indicated in the apparatus diagram.

Because the apparatus could only measure the upward moving β particles, the experimental determination of the discrepancy between upward and downward moving particles had to be conducted in two stages. First, the Co^{60} was polarized

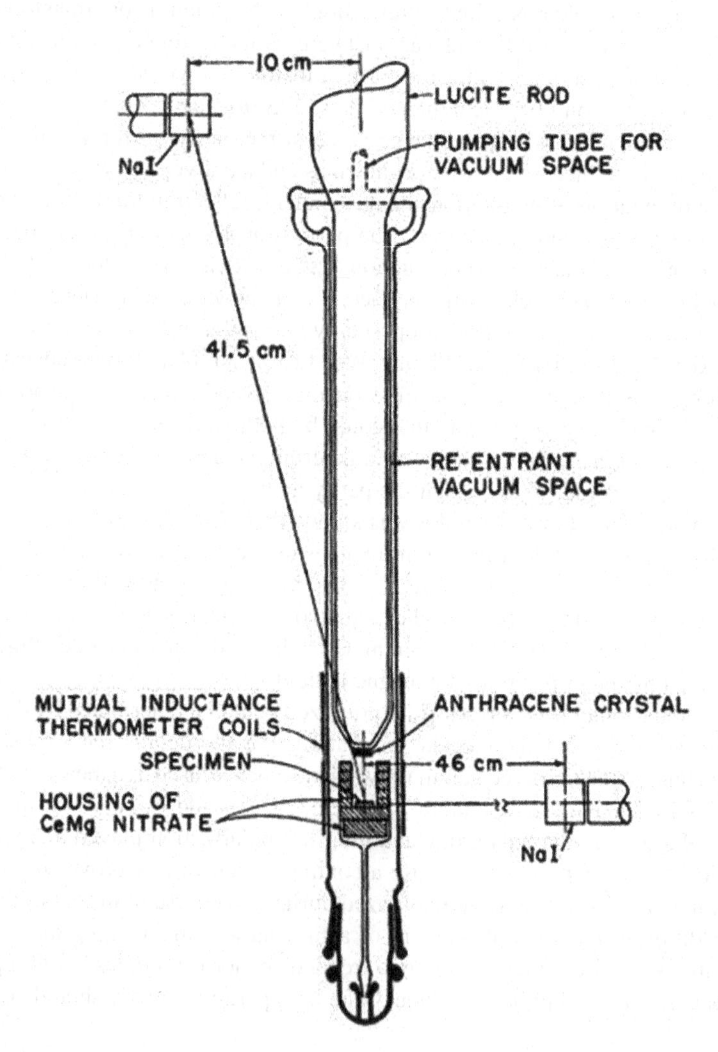

Fig. 7.3 The experimental apparatus of Wu et al. *Source* Wu et al. (1957)

after which the "beta and gamma counting was then started (p. 1414)." This configuration though only took account of the beta electrons that were moving upwards. To determine the count of the downward moving beta electrons the Co^{60} had to be repolarized by changing the magnetic field to the opposite direction. In effect this turned the Co^{60} specimen upside down with respect to the anthracene crystal so that the downward moving beta particles could be registered.

The two parts of the actual experiment are depicted at the left side of Fig. 7.4, while the right side represents what the experiment would have looked like if the apparatus could have detected the downward moving beta particles. Converting the real experiment into the counterfactual description involves flipping the upward moving beta particles of the second part of the real experiment upside down which is equivalent to rotating the real experiment 180° around the in-and-out of page y axis. Thus, for example, an observer viewing the second part of the actual experiment from above would "see" approaching beta particles with a counterclockwise magnetic field. Similarly, if viewing the rotated (counterfactual) experiment from beneath the observer would again see approaching beta particles with a counterclockwise spin. We should note that the counterfactual configuration (the right side of Fig. 7.4) is commonly used in textbook and other secondary accounts of the experiment without comment as to its ancestry.

The collected data on the beta and gamma counts were presented in the form as shown in Fig. 7.5. The gamma anisotropy is displayed in the top graph for the two different polarizations of the Co^{60}. As the apparatus grew warmer the polarization decreased as shown by the fact that the gamma anisotropy values converged to the

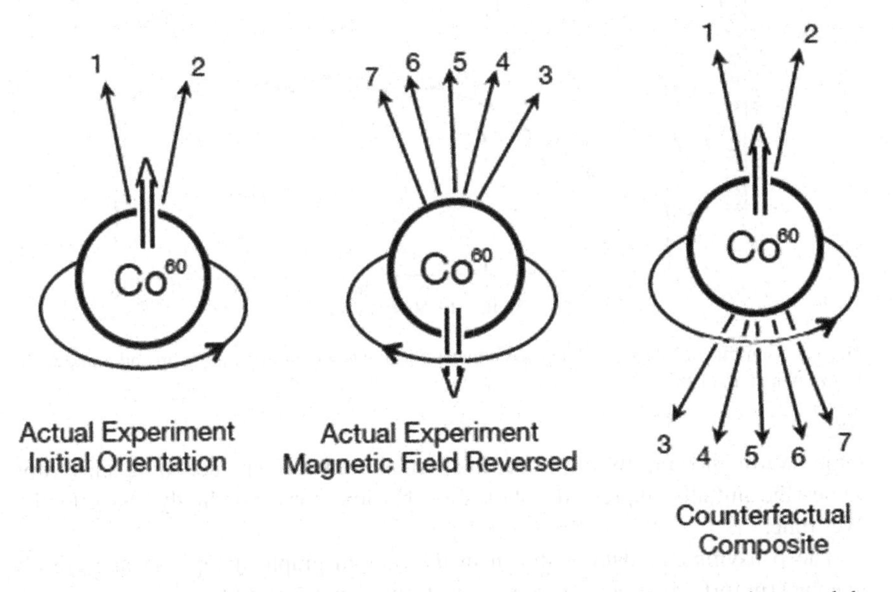

Fig. 7.4 Schematic representation of the two parts of the actual Wu et al. experiment and the counterfactual equivalent. The spin is represented by the double arrows pointing up or down. *Source* Authors

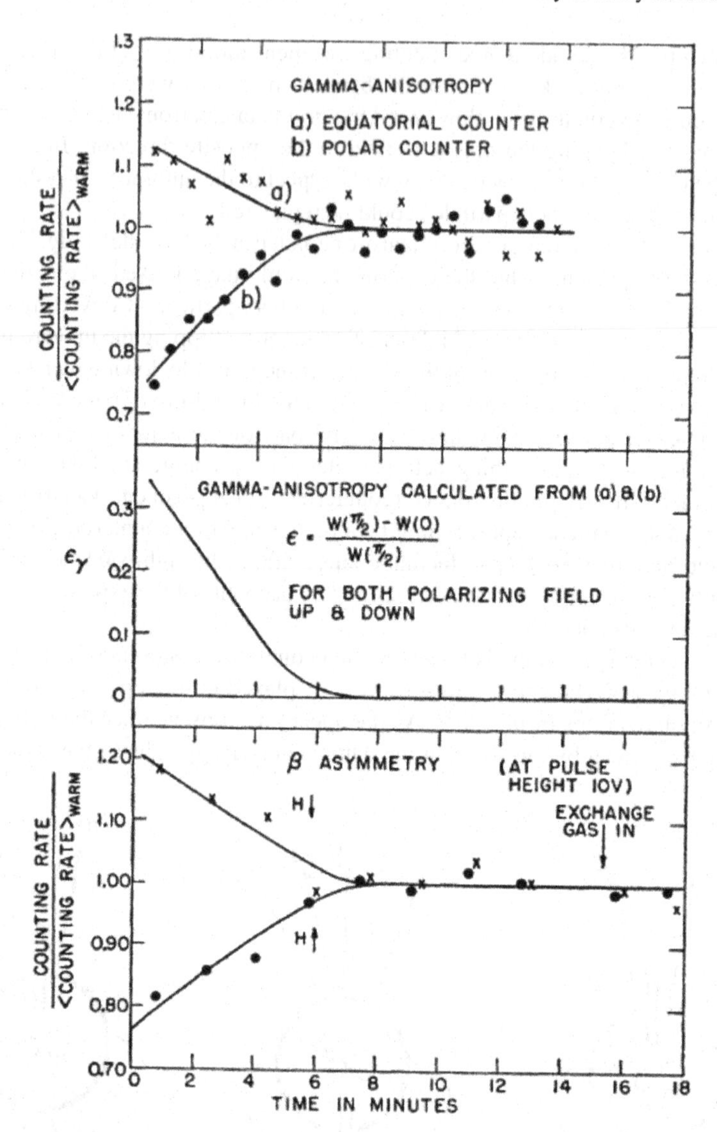

Fig. 7.5 Gamma anisotropy and beta asymmetry for polarizing field pointing up and point down. *Source* Wu et al. (1957)

sable "warm" nonanisotropic value. In short, as shown by the declining anisotropy values the initially impressed polarization became correspondingly less effective over time.

The β asymmetry data is shown in the bottom graph of Fig. 7.5 and closely matched (in form) that of the gamma anisotropy—which must have been an exceedingly gratifying result for Wu and her experimental cohort since it indicated that the β

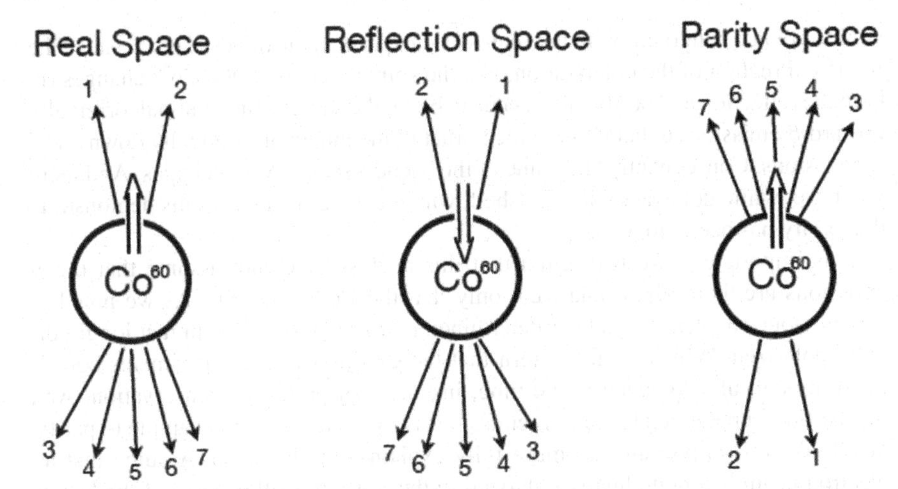

Fig. 7.6 Schematic representation of the Wu et al. experiment showing the "Real Space" result and the parity "Mirror Space" counterpart showing the change in the relative orientation of the asymmetrical data with respect to the Co^{60} spin *Source* Authors

asymmetry varied as the polarization of the Co^{60} and not with respect to some varying confounding influence.[14] The top curve shows the ratio of the downward moving β electrons ejected from polarized Co^{60} as compared with β electrons ejected from unpolarized Co^{60}. Similarly, the bottom curve shows the asymmetry for the upward moving β electrons. Bottom line: Maximum emission occurs in a direction contrary to that of the polarization of the Co^{60}.

> The sign of the asymmetry coefficient . . . is negative, that is, the emission of beta particles is more favored in the direction opposite to that of the nuclear spin (Wu et al. 1957, p. 1414).

Question: How do we get from this experimental result to the higher-level theoretical result that parity has been violated? To see how the argument goes see Fig. 7.6 which represents in schematic form the essential features of the Wu experiment and its relation with the reflection and parity coordinate transformations. In order to highlight how these transformations apply we have taken the liberty of labeling the paths of the β emissions. Such identification, of course, was not available to Wu and her colleagues since only the aggregate counts for the β emissions were experimentally accessible. Nevertheless, as can be readily seen, the reflection and parity mirror

[14] It should be noted that there was more involved here than just making a visual comparison of these two sets of results as shown by the fact that Wu and her cohort calculated the resultant anisotropy from the two value sets displayed at the top of Fig. 7.4 and presented this result in the graph at the middle of that figure. And because gamma radiation is the result of an electromagnetic process its production satisfies parity which means that the resultant gamma anisotropy provided a reliable background standard which in turn was used to determine the *amount* of parity violation (the asymmetry coefficient) exhibited in beta decay. We'll return to this determination in the concluding section of this chapter where we consider its relevance in determining whether once was enough for this experiment to have established parity nonconservation.

worlds are *not* congruent with the real-world experimental values. This because the *relative* direction of the polarization, i.e., the spin direction, of the Co^{60} changes (in both the reflection and parity mirrors) from being the same as the minimal, originally upward β emissions to being the same as that of the maximal (originally downward) β emissions. Consequently, the value of the pseudo-scalar ($\boldsymbol{S}*\boldsymbol{p}$) changes. And there you have, while not exactly in a nutshell, why the experimental results demonstrate that parity has been violated.

It is important to keep in mind that this analysis does not assume that the β emissions are themselves polarized, only that the Co^{60} source is. As we noted in our previous chapter, it was later determined that β emissions are in fact longitudinally polarized. Which meant in turn that the Cox and Chase experiments constituted, though unrecognized at the time, instances of parity nonconservation. And this because a polarized beam of β emissions is *by itself* a counterexample to parity. We'll take this opportunity to more fully explain why this is so. Assume that an electron beam is longitudinally polarized in the sense that all or most of the beams consists of electrons as configured on the left side (the "Real Space") of Fig. 7.7. The parity mirror image is pictured on the right where the crucial difference is that the spin is now in the same direction as the electron momentum—which is a violation of parity. The only way to avoid this conclusion would be if for every electron of the configuration on the left side of Fig. 7.7 there was a corresponding electron *in the same beam* of the configuration as on the right side of the figure. But this is impossible given that the beam is polarized.

So in the case of a longitudinally polarized beam of β emissions parity is violated in very much the same way that it was violated in the case of the Wu experiment, namely, that spin direction remains intact after the parity transformation while all else changes. Consequently, in the case of the Wu experiment, the spin orientation changes with respect to the transformed asymmetry, while in the case of a longitudinally polarized beta particle beam the spin orientation changes with respect to the transformed momentum.

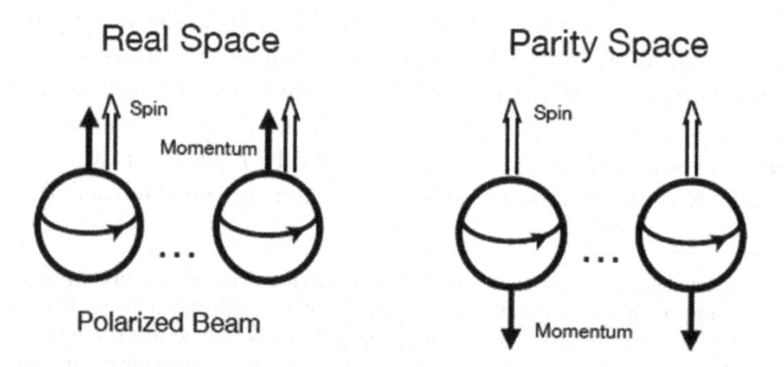

Fig. 7.7 "Real Space" longitudinally polarized bean and its parity "Mirror Space" counterpart showing the change in the relative orientation of the electron momentum with respect to spin. *Source* Authors

As was the case in our earlier case studies, there was considerable discussion by Wu and her colleagues regarding systematic uncertainty. So, for example, the fact that the warm counting rates (i.e. for no polarization) were independent of the polarizing field direction argues against any significant instrumental asymmetry. Similarly, for the concordance between the β asymmetry and the gamma anisotropy which also persisted despite the change in polarizing field. It was also possible that the demagnetization field used to cool the sample might have left a remnant magnetization that caused the β-ray asymmetry. This confounding possibility was eliminated by noting that the observed asymmetry did not change sign with the reversal of the field direction (Wu et al. 1957, p. 1414).

A last systematic uncertainty that had to be dealt with was that there might have been a small magnetic field perpendicular to the polarizing field due to the fact that the Co^{60} crystal axis was not parallel to the polarizing field. Eliminating this possibility and a related companion uncertainty was somewhat more involved:

> To check whether the beta asymmetry could be caused by such a magnetic field distortion, we allowed a drop of CoCl solution to dry on a thin plastic disk and cemented the disk to the bottom of the same housing. In this way the cobalt nuclei should not be cooled sufficiently to produce an appreciable polarization, whereas the housing will behave as before. The large beta asymmetry was not observed. Furthermore, to investigate possible internal magnetic effects on the paths of the electrons as they find their way to the surface of the crystal, we prepared another source by rubbing CoCl on the surface of the cooling salt until a reasonable amount of the crystal was dissolved. We then allowed the solution to dry. No beta asymmetry was observed with this specimen (Wu et al. 1957, p. 1414).

In sum, while the experiment was relatively straightforward in its general concept it was extremely challenging to bring that concept to life in the form of a real-world experiment that would be not only credible but decisive. We'll return to the question of whether once was enough to establish parity nonconservation later, but for now we'll move on to a brief consideration two other experiments that demonstrated parity nonconservation.

7.3 The Sequential Decay $\pi \to \mu \to e$

In addition to the beta decay experiment successfully executed by Wu, Lee and Yang also suggested the following decay process as a promising candidate for experimental examination: $\pi \to \mu + \nu$, $\mu \to e + \nu + \nu$ (Lee and Yang 1956, p. 257). The basic idea here was that one could compare the muon spin (before the electron decay) with the anticipated asymmetrical distribution of the electron decay and thereby have an identified orientation of spin and asymmetry that was not preserved in the parity mirror image of the experiment. As stated by Lee and Yang:

> If parity is conserved in neither [of these processes] the distribution will not in general be identical for θ and $\pi - \theta$. To understand this, consider first the orientation of the muon spin. If [the first process] violates parity conservation, the muon would be in general polarized in its direction of motion. In the subsequent decay [the second process], the angular distribution

problem with respect to θ is therefore closely similar to the angular distribution problem of β rays from oriented nuclei, which we have discussed before (Lee and Yang 1956, 257).

This proposal was successfully executed in Garwin et al. (1957) and using a different technique in Friedman and Telegdi (1957). We'll begin with the Garwin experiment where the apparatus used is shown in Fig. 7.8. The central fact that makes the experiment possible is that the 85 MeV pion beam from the Nevis cyclotron was known to "contain about 10% of muons which originate principally in the vicinity of the cyclotron target by pion decay-in-fight" (Garwin et al. 1957, p. 1415). This meant that a carbon absorber could be used to absorb the pi mesons but that at the same time would allow the muons to pass through, where they would eventually stop in the carbon target and decay into electrons which then made their way to the detecting counters.

The analysis of the experiment in terms of the underlying processes involved is somewhat unusual in that Garwin begins with a hypothesis about what's involved and

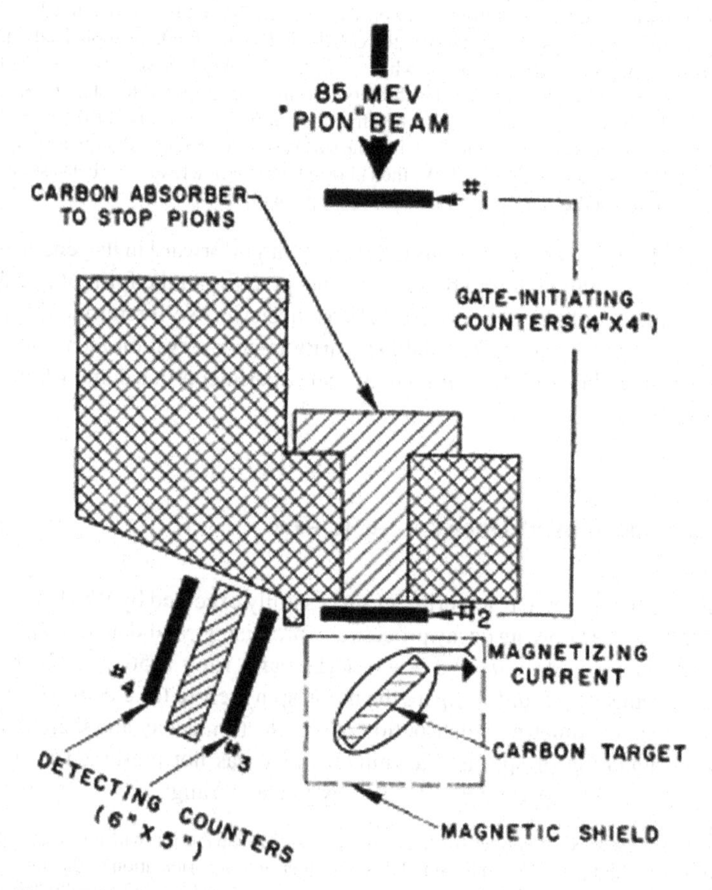

Fig. 7.8 The experimental apparatus of Garwin et al. *Source* Garwin et al. (1957)

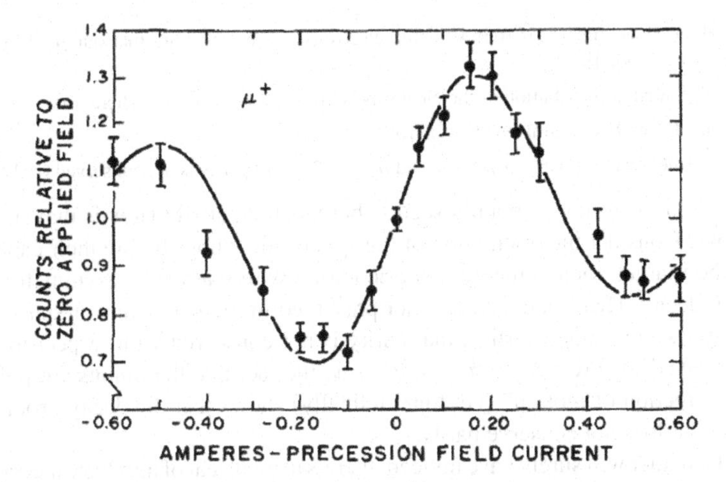

Fig. 7.9 Variation of counting rate with magnetizing current. The solid curve is computed from an assumed electron angular distribution (1–1/3 cosθ). *Source* Garwin et al. (1957)

then shows that the hypothesis is confirmed by the experimental results. Thus we are asked to consider the "possibility" that the muons are created "with large polarization in the direction of motion" and to "assume" that "the processes of slowing down, stopping, and the microsecond of waiting do not depolarize the muons" (p. 1416).[15] If so, then "the electrons emitted from the target may have an angular asymmetry about the polarization direction … of the form $1 + \alpha \cos\theta$" (p. 1416).

But even if all this were true the detecting counters could only register the electrons that came in at a fixed angle with respect to the apparatus. To circumvent this shortcoming, "a small vertical field" was applied to the muons (after their escape from the carbon absorber) which caused them to rotate, which in turn ultimately caused the dispersion pattern of the electrons to rotate as well. In this way the anticipated dispersion asymmetry could be detected.

A sample of the data runs is shown in Fig. 7.9. Here as can be readily seen the smaller variation from the base value (at zero applied vertical field) is around 0.15 (in electron counts), while the larger value is around 0.3. The asymmetry therefore is roughly $(0.3 - 0.15/0.3 + 0.15) = 0.33$. This result serves two purposes. First, it confirms the hypothesis proposed about the operation of the experiment We should add that confirmation here assumes the absence of any reasonable alternative. And second that both decay processes violate parity conservation. As stated more fully by Garwin the experiment "establishes the following facts"[16]:

[15]The "microsecond of waiting" was determined by the gate-initiating counters which created an opening gate of duration 1.25 μs (Garwin et al. 1957, p. 1415). This was to guard against muon decays in flight.

[16]There are seven more "facts" that are claimed to follow from the experimentally determined results which deal with charge conjugation, spin, magnetic moment and other matters of interest. But since our focus here is on parity nonconservation we will not pursue this bounty of additional results.

I. A large asymmetry is found for the electrons . . . establishing that our μ^+ beam is strongly polarized.

II. The angular distribution of the electrons is given by $1 + \alpha\, cos\, \theta$, where α is

  a. 1/3 with an estimated error of 10%.

III. In [the] reactions [$\pi^+ \rightarrow \mu^+ + \nu$ and $\mu^+ \rightarrow e^+ + 2\nu$] parity is not conserved (p. 1415).

That I and II follow is readily seen[17], but what about III? How is that supposed to follow? Consider the orientation of the spin of the muons before their capture by the carbon target, then compare this orientation with that of the asymmetry of the electron decay. That relationship is not preserved in the parity mirror space which means (as we explained earlier) that parity is not conserved with respect to $\mu^+ \rightarrow e^+ + 2\nu$. And with respect to $\pi^+ \rightarrow \mu^+ + \nu$, the fact that the muons are polarized "in the direction of motion" (i.e. longitudinally) shows that the decay process that produced it does not conserve parity.

As the reader will surely have noticed, there's a great deal of auxiliary theory that's needed to tease out the nonconservation result reported in the Garwin experiment. Certainly significantly more than required in the Wu experiment. Consequently, there was much more to worry about with respect to systematic uncertainty. Which suggests that when it comes to determining whether once was enough the Wu experiment will be the stronger candidate. In this regard it is significant though not unexpected that during the course of the Garwin experiment various "systematic checks" and error determinations were made. See (Garwin et al. 1957, p. 1416 and p. 1417 n 10).

The same decay process was examined in Friedman and Telegdi (1957) but this time using nuclear emulsions to separate out the decay components. Here the idea was "to observe the pertinent correlation by bringing the π^+ mesons to rest in a nuclear emulsion in which the μ^+ meson also stops" and where the "pertinent correlation" was the relationship between "the initial direction of motion of the muon in the process" and the corresponding asymmetry in the electron decay (p. 1681). As further explained, "we exposed ... nuclear emulsion pellicles ... to a π^+ beam of the University of Chicago synchrocyclotron" where "the pellicles were contained inside three concentric tubular magnetic shields and subject to 4×10^{-3} gauss" (p. 1681). The shields guarded against any stray magnetic fields which might mask the predicted effect.

And while no additional experimental details were given it was confidently stated that with respect to confounding causes "[o]ne has only to bear in mind two facts":

> (1) even weak magnetic fields, such as the fringing field of a cyclotron, can obliterate a real effect . . . (2) μ^+ can form "muonium, " i.e., (μ^+e^-), and the formation of this atom can be an additional source of depolarization . . . ." (p. 1681).

But there was no account given as to how these complications were to be dealt with other than noting that "[i]n the absence of specific experiments on muonium formation, one can perhaps be guided by analogous data on positronium in solids

---

[17]The first result (I) does require some auxiliary theory and information about how the asymmetry in the electron distribution shows that the μ^+ beam must have been polarized as a consequence of the first process, i.e., $\pi^+ \rightarrow \mu^+ + \nu$.

(p. 1681)." Nor was there any indication as to how the measurements of the asymmetry were affected with only the final value of the lower limit of the asymmetry as 0.062 ± 0.027 being given (later 0.91 ± 0.027.[18] By way of explanation for the paucity of relevant details Friedman and Telegdi remarked that "[i]n view of the intrinsic importance of the subject, we consider it worthwhile to present our data at this preliminary stage" (1682).

That there was some haste involved in all this is further indicated by the fact that there was a note added in proof stating: "From 2000 events [i.e., 700 plus the initial 1300], we get for this ratio 0.091 ± 0.022"—which was a remarkable improvement based on just 700 additional events. For some well-considered skepticism about the experiment see (Goudsmit 1971) who was at the time the editor of *Physical Review*. Bottom line: Not nearly as good a candidate for once being enough as the Wu and Garwin experiments.

7.4 The Historical Judgment: Once Was Enough

The three experiments, those of Wu and her collaborators along with those of Friedman and Telegdi (1957) and of Garwin et al. (1957) persuaded the physics community that parity was not conserved in the weak interaction.

Perhaps the strongest evidence supporting the view that these were crucial experiments is the fact that Lee and Yang were awarded the Nobel Prize in physics within a year of the publication of their paper proposing parity violation. (In the intervening time the experimental results demonstrating parity nonconservation had been published). This is the shortest time between an accomplishment and an award of a Nobel Prize in history. The prize presentation speech of O.B. Klein makes the impact of this work clear. "In fact, most of us were inclined to regard the symmetry of elementary particles with respect to left and right as a necessary consequence of the general principle of right-left symmetry of Nature. Thanks to Lee and Yang and the experimental discoveries inspired by them we now know that this was a mistake… nor could I do justice to the enthusiasm your new achievement has aroused among physicists. Through your consistent and unprejudiced thinking you have been able to break a most puzzling deadlock in the field of elementary particle physics where now experimental and theoretical work is pouring forth as a result of your brilliant achievement (Klein 1964, p. 390)."

The work Klein referred to supports his view. A survey of the literature at the time indicates that there were no attempts to save parity nonconservation in the weak interactions involving neutrinos.[19] The discovery of parity nonconservation had a dramatic effect on the work done by physicists. There were a significant number of

[18] We should note, however, that there had been considerable prior use of nuclear emulsions to study high-energy particles. This began with the work of Marietta Blau and Hertha Wambacher in 1937. We also note that the effects discussed would mask any possible asymmetry, not create one.

[19] There are, in fact, some non-leptonic weak decay amplitudes that do conserve parity. An amusing and interesting feature of several papers on such decays is that they dealt with the $\theta-\tau$ puzzle, which

additional experiments reported that confirmed the initial results.[20] Sullivan et al. (1977) reported a large increase in the number of theoretical articles published in the field of weak interactions in 1957 and 1958. There was no similar effect seen in experimental articles, probably due to the need for time to construct experimental apparatuses. If, however, we look at the numbers of experimental papers devoted to various topics in weak interactions, we find that in both 1958 and 1959 approximately 60 percent of the articles were devoted to parity and its closely related field of the V – A theory of weak interactions.[21] White and his collaborators also show that there were large increases in the number of both experimentalists and theorists entering the field of weak interactions in 1957 and 1958, as shown by the dates of their first publications.

Textbooks tell us what is accepted as knowledge by the physics community at a given time so it is interesting to look at accounts of parity conservation in texts published before and after the discovery. A typical prediscovery statement is in Halliday's *Introductory Nuclear Physics*, a textbook that was in wide use during the 1950s. In particular, it was the text used in Wu's graduate course on nuclear physics given at Columbia in the spring of 1959.[22] "Parity is an interesting property because of the law of conservation of parity. This states that the parity of an isolated system cannot change, no matter what transformations or recombinations take place within it (Halliday 1955, p. 38)." Needless to say, Professor Wu corrected the text statement and discussed the nonconservation of parity in class. One can also look at two different editions of the same text published before and after the 1957 discovery. In the second edition of Schiff's *Quantum Mechanics*, a text widely used in graduate courses at the time, Schiff remarks, "All the interactions between particles thus far encountered in physical theory lead to Hamiltonian functions that are unchanged by reflection [i.e., parity is conserved] (Schiff 1955, p. 160)." In the third edition of the book, he states, "With the exception of the weak interactions, which among other things is responsible for radioactive beta decay, all known Hamiltonians commute with U_r [conserve parity] (Schiff 1968, p. 254)."

There is also anecdotal evidence supporting the view that these experiments were indeed crucial. Consider the history of comments made by Wolfgang Pauli, a Nobel Prize winner. In 1933 Pauli rejected Weyl's two-component theory of spin ½ particles (Pauli 1933). He pointed out that Weyl's theory gave a mass of zero for the particles and was not invariant under space reflection. "However, as the derivation shows, these wave equations are not invariant under reflection (interchanging left and right) and thus are not applicable to physical reality (p. 226)." As late as 17 January 1957, Pauli had not changed his mind. In a letter to Victor Weiskopf, he wrote, "I do not

was the problem that stimulated Lee and Yang to suggest parity nonconservation. For details see Franklin (1986, Footnote 110, p, 252).

[20]For details see Franklin (1986, Footnote 51, pp. 249–250).

[21]The V – A theory was proposed to incorporate parity nonconservation into the theory of weak interactions.

[22]Franklin was a student in Professor Wu's graduate course in nuclear physics during the spring of 1959.

believe that the Lord is a weak left-hander, and I am willing to bet a very large sum that the experiments will give symmetric results (Bernstein 1967, p. 59)."[23] When Pauli received preprints of the experimental papers he accepted the results and their significance immediately. In another letter to Weiskopf, he wrote, "Now, after the first shock is over, I begin to collect myself. Yes, it was very dramatic. On Monday, the twenty-first, at 8 p.m. I was to give a lecture on the neutrino theory. At 5 p.m. I received the three experimental papers [those of Wu, Garwin, and Telegdi]. I am shocked not so much by the fact that the Lord prefers the left hand as by the fact that He appears to be left-right symmetric when he expresses Himself strongly. In short, the actual problem now seems to be the question: Why are strong interactions right and left symmetric (Bernstein 1967, p. 60)."

Pauli was not alone in his willingness to bet against parity nonconservation. Richard Feynman bet Norman Ramsey, both Nobel Prize winners, $50 to $1 that parity would be conserved. Feynman paid.[24] T.D. Lee also reported that Felix Bloch, yet another Nobel Prize winner, offered to bet other members of the Stanford Physics Department his hat that parity was conserved. He later remarked to Lee that it was fortunate he didn't own a hat (Lee 1985).

The evidence seems clear that the physics community regarded the experiments as decisive. They established the nonconservation of parity.

7.5 Conclusion

Here's the question we raised earlier but did not answer: Was once enough for the Wu experiment to have decisively demonstrated that parity conservation was violated in the case of β-decay? Allow us to refocus the question. Was the reported data, even if accepted, good enough for the purpose of showing that parity conservation was violated? In order to satisfy this purpose the asymmetry data was *by itself not sufficient*. This because (1) what may be roughly described as the inherent asymmetry revealed by the γ anisotropy had to be taken into account; (2) the relativistic term v/c had to be determined and taken into account; and (3) corrections still had to be made for "backscattering" of the electrons. Wu did not go into detail about how these items were taken into account and gave only the following summary.

> The exact evaluation of [the asymmetry coefficient] α is difficult because of the many effects involved. The lower limit of α can be *estimated roughly*, however, from the observed value of asymmetry corrected for backscattering. At velocity $v/c \approx 0.6$, the value of α is *about* 0.4. The value of $\langle Iz \rangle / I$ can be calculated from the observed anisotropy of the gamma radiation to be *about* 0.6. These two quantities give the lower limit of the asymmetry parameter $\beta(\alpha = \beta\langle Iz \rangle / I)$ *approximately* equal to 0.7 (Wu et al. 1957, p. 1414, emphasis added).[25]

[23] Pauli was fortunate that he did not bet that large sum of money.

[24] Private communication to Franklin from Norman Ramsey, who gave him a copy of a letter he sent to Feynman recounting that story. The story also appears in Frauenfelder and Henley (1975, p. 389) and in Feynman (1985, p. 248)

[25] For more details see Wu's more expansive account in Wu (2008, p. 59).

Despite the absence of detail, what is clear is that:

> In order to evaluate α, accurately, many supplementary experiments must be carried out to determine the various correction factors. It is estimated here *only to show the large asymmetry effect* (p. 1414, emphasis added).

It appears that once was not enough to achieve anything more than a rough approximation of the asymmetry coefficient. Nevertheless, because the estimated value (of the lower limit) was so large, it was more than enough to decisively demonstrate parity nonconservation. For unlike the case of a null experiment (such as the demonstration of the equality of gravitational and inertial mass) where increasingly greater accuracy is required, such excruciating precision was not required here.[26] And because such precision was not required, Wu did not bother to provide any statistical analysis of the counting error involved in the determination of the asymmetry values reported in Fig. 7.4. To do so would have been pointless given the obviously inconsequential error estimate that would have resulted.

While our argument is that the Wu experiment *should* have been deemed enough for showing the violation of parity in the case of beta decay, it was clearly deemed by the scientific community *to have done so*—though perhaps with the assistance of the (Garwin et al. 1957; Friedman and Telegdi 1957) experiments. And having been deemed to have done so the effect, in terms of pursuit, was that research on parity "advanced at an unprecedented pace."

> First, the non-conservation of parity was also observed in the $\pi^{\pm} \rightarrow \mu^{\pm} \rightarrow e^{\pm}$ decays and other weak interactions *not restricted to nuclear beta-decays*. Thus the parity non-conservation is a *fundamental characteristic of the weak interactions*. ... The (v/c) dependence of the asymmetry parameter A of the beta particles from the polarized ^{60}Co was also used to examine the validity of the Time Reversal "T" and it was found, in general, sound (Wu 2008, p. 61, emphasis added).

These successful applications and extensions of the Wu experiment served to *indirectly replicate* the experiment in the sense of providing evidence that it had been soundly conducted. With regard to further pursuit, Wu did not rest on her laurels and was indefatigable in her efforts "to improve the cryogenic condition and magnetic field shielding of the parity experiments so that more reliable and precision results could be derived from asymmetry measurements (p. 61)."[27] Included in these efforts were the determination of certain Isospin-Hindered β-Transitions that were "of particular interest for studies of isospin conservation of nuclear forces and the Time-Reversal Invariance (TRI) tests of the weak interactions (p. 64)."

Historically, as noted above, there was more than just the Wu experiment to thank for these developments, namely, Garwin et al. (1957) and Friedman and Telegdi (1957). So the question of whether and for what purpose once was enough for the Wu experiment itself was not directly confronted by the physics community and did not need to be. Since the trio of experiments was more than enough there was no reason for the scientific community to consider the question.

---

[26] For many such examples and the methodological issues involved see (Franklin and Laymon 2019).

[27] For a listing and detailed description of these improvements see Wu (2008, pp. 61–64).

Still, however misguided it may appear, we think the counterfactual question of what the judgment would have been had the Garwin and Friedman experiments not been successfully completed is worth considering. This because such an exercise serves to highlight the *normative values* in play. But we have already conducted such an exercise insofar as our analysis above of the Wu experiment is in large part normative and thus provides evidence that in the counterfactual situation considered once would have been enough. Though we hasten to add that our determination of the normative values in play ultimately draws its inspiration and support from the historical decisions made by the scientific community. Thus, we have, in other words, taken for granted a sophisticated form of *crowd sourcing* as the foundation for our normative standards.

Viewing the epistemology of experimentation this way does serve to remind us of the fact that there will be variations in the determinations by individual scientists as to whether once was enough. With respect to questions of pursuit answers will depend, for example, on the perceived promise of the undertaking, the skill set and career possibilities of the scientist considering what to pursue. And if the question is one involving the allocation of resources among competing interests and scientists, answering the question becomes even more involved. The question of whether the Wu experiment was enough to decisively refute parity nonconservation seems, by comparison, relatively clear cut. Still that decision will crucially depend on the suspicions and anticipations of the individual scientist about the existence of accidental and otherwise unaccounted for confounding influences. Thus, for example, when Pauli was first informed of the Wu experiment showing that parity was not conserved, Pauli responded: "That's total nonsense." And when told that "the experiment says it is not," Pauli curtly parried: "Then it must be repeated."[28] So much for once being enough! But fortunately for us, Pauli soon changed his mind on this.

References

Alford, W.L. and R.B. Leighton. 1953. Mean lifetimes of V-particles and heavy mesons. *Physical Review* 90: 622–629.
Ambler, E., M.A. Grace, et al. 1953. Nuclear polarization of Cobalt 60. *The London, Edinburgh, and Dublin Philosophical Magazine and Journal of Science* 44: 216–218.
Ballam, J., V.L. Fitch, T. Fulton, K. Huang, R.R. Rau, and S.B. Treiman. 1956. High Energy Nuclear Physics. In *Sixth Annual Rochester Conference, Rochester, Interscience.*
Bernstein, J. 1967. *A comprehensible world*. New York: Random House.
Feynman, R.P. 1985. *Surely you're joking, Mr. Feynman*. New York: Norton.
Franklin, A. 1986. *The neglect of experiment*. Cambridge, Cambridge University Press.
Franklin, A., and R. Laymon. 2019. *Measuring nothing, repeatedly*. San Rafael, CA: Morgan and Claypool.
Frauenfelder, H., and E. Henley. 1975. *Nuclear and particle physics*. Reading, MA: W.A. Benjamin.
Friedman, J.L., and V.L. Telegdi. 1957. Nuclear emulsion evidence for parity nonconservation in the decay chain $\pi^+ - \mu^+ - e^+$. *Physical Review* 105: 1681–1682.

[28] Hudson (2001).

Garwin, R.L., L.M. Lederman, et al. 1957. Observation of the failure of conservation of parity and charge conjugation in meson decays: The magnetic moment of the free muon. *Physical Review* 105: 1415–1417.

Gell-Mann, M. 1953. Some remarks on the V-particles. *Physical Review* 92: 833–834.

Gell-Mann, M. 1956. The interpretation of the new particles as displaced charge multiplets. *Nuovo Cimento* (Suppl) 4: 848–866.

Gibson, W.M., and B.R. Pollard. 1976. *Symmetry principles in elementary particle physics.* Cambridge: Cambridge University Press.

Goudsmit, S.A. 1971. A reply from the editor of physical review. *Adventures in Experimental Physics Gamma*: 137.

Halliday, D. 1955. *Introductory nuclear physics.* New York: Wiley.

Hudson, R.P. 2001. Reversal of the parity conservation law in nuclear physics. In *A century of excellence in measurements, standards, and technology NIST special publication*, ed. D.R. Lide, vol. 958, 111–115. National Institute of Standards and Technology.

Jordan, P., and R.D. Kronig. 1927. Movement of the lower jaw of cattle during mastication. *Nature* 120: 807.

Klein, O.B. 1964. Nobel prize presentation speech. In *Nobel lectures, physics, 1942-1962*, vol. 390. Amsterdam: Elsevier.

Laporte, O. 1924. Die Struktur des Eisenspektrums. *Zeitschrift fur Physik* 23: 133–175.

Lee, T.D. 1971. *History of weak interactions.* New York: Columbia University.

Lee, T.D. 1985. Letter. A. Franklin.

Lee, T.D., and J. Orear. 1955. Speculations of heavy mesons. *Physical Review* 100: 932–933.

Lee, T.D., and C.N. Yang. 1956. Question of parity nonconservation in weak interactions. *Physical Review* 104: 254–258.

Maglic, B. (ed.). 1973. *Adventures in experimental physics.* Princeton: World Science Education.

Nakano, T., and K. Nishijima. 1953. Charge independence for V particles. *Progress in Theoretical Physics* 10: 581–582.

Nishijima, K. 1954. Some remarks on the even-odd rule. *Progress in Theoretical Physics* 12: 107–108.

Nishijima, K. 1955. Charge independence theory of V particles. *Progress in Theoretical Physics* 13: 285–304.

Orear, J., G. Harris, et al. 1956. Spin and parity analysis of Bevatron τ mesons. *Physical Review* 102: 1676–1684.

Pais, A. 1951. Some remarks on the V-particles. *Physical Review* 86: 663–672.

Pauli, W. 1933. Die Allgemeinen Prinzipen der Wellenmechanik. *Handbuch der Physik* 24: 83–272.

Ramsey, N. 1956. *Molecular beams.* Oxford: Oxford University Press.

Rochester, G.D., and C.C. Butler. 1947. Evidence for the existence of new unstable elementary particles. *Nature* 160: 855–857.

Schiff, L.I. 1955. *Quantum mechanics.* New York: McGraw Hill.

Schiff, L.I. 1968. *Quantum mechanics.* New York: McGraw Hill.

Sullivan, D., D.H. White, et al. 1977. The state of science: Indicators in the specialty of weak interactions. *Social Studies of Science* 7: 167–200.

Wick, G.C., A.S. Wightman, et al. 1952. The intrinsic parity of elementary particles. *Physical Review* 88 (1): 101–105.

Wigner, E. 1927. Einige Folgerungen aus der Schrodingerschen Theorie fur die Termstrukturen. *Zeitschrift fur Physik* 43: 624–652.

Wu, C.S. 2008. The discovery of the parity violation in weak interactions and its recent developments. *Lecture Notes in Physics* 746: 43–69.

Wu, C.S., E. Ambler, et al. 1957. Experimental test of parity nonconservation in beta decay. *Physical Review* 105: 1413–1415.

Yang, C.N. 1956. Theoretical interpretation of new particles. In *Sixth Annual Rochester Conference, Interscience.*

Part III
Once Wasn't Enough

Chapter 8
The Search for the Magnetic Monopole

Dirac was frugal when it came to managing mathematics. He was not one to let lie idle what were otherwise thought to be mere accidental and inconsequential fragments. Thus, as we have seen in an earlier chapter, he recruited the negative solutions to his eponymous equation to serve as the foundation for a new particle, the positron. So too with the magnetic monopole. Here the mathematical flotsam, as it were, consisted of a set of equations that were symmetrical twins of the usual Maxwell equations but where the twins featured magnetic monopoles as the formal equivalents of the more ordinary positive and negatively charged particles.

This formal equivalence was known by the late nineteenth century but other than Pierre Curie no one thought that there might be ontological consequences—that monopoles might exist as fundamental particles on par with electrons and protons. Dirac put a stop to such waste. We will briefly review how he did so. But our principal focus will be how one experimental team—though there were many others that tried—developed the superior apparatus and experiment that they hoped would capture what had been the hitherto elusive magnetic monopole.

8.1 Magnetic Monopoles

Ordinary magnets have two poles, north and south. If you cut a magnet in half, you don't get two separate poles but rather two smaller magnets each with a north and a south pole. It's not as if the north pole had its tail cut off and thereby establishes some sort of independent existence that lasts until connection is made with the other end of the now divided magnet. Such magnetism is the result of the electron orbits and spin having been oriented so that their individual magnetic fields line up. Thus, the tail is always present.

What then is a magnetic monopole? Maxwell's equations govern the relationship between moving charged particles and their accompanying magnetic field. But these equations have a symmetric set of mathematical twins, a sort of mirror image, where

A. Franklin and R. Laymon, *Once Can Be Enough*,
https://doi.org/10.1007/978-3-030-62565-8_8

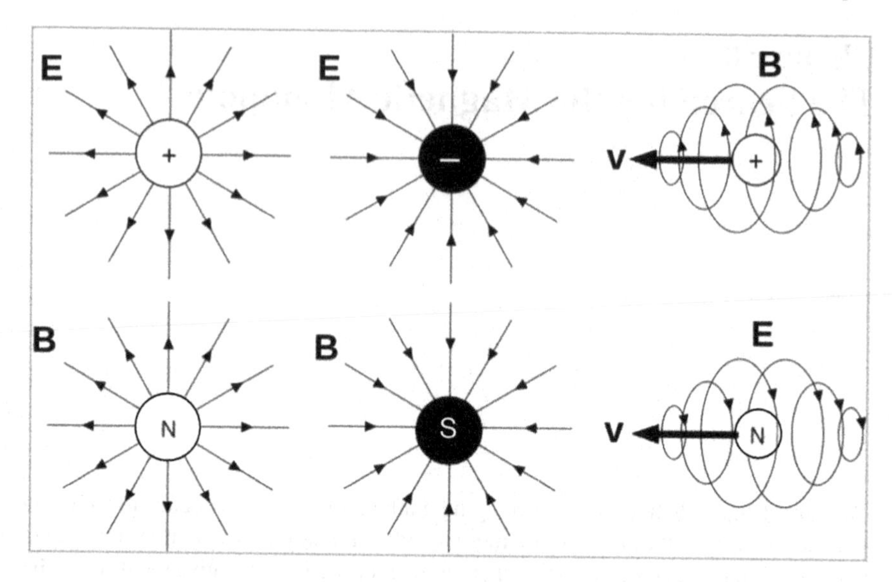

Fig. 8.1 Left: Fields due to stationary electric and magnetic monopoles. Right: In motion (velocity v), an *electric* charge induces a B field while a *magnetic* charge induces an E field. Conventional current is used

the electric and magnetic fields reverse roles whereby a moving magnetic monopole is now accompanied by an electric field. Thus, a "north" magnetic monopole corresponds to a proton and a "south" corresponds to an electron. For a clear and striking graphical representation of this symmetry in action see Fig. 8.1 which makes clear the equivalent formal status of magnetic monopoles.[1] Consequently a moving monopole, if it existed, would create an electric field just as a moving charge creates a magnetic field. The fact that Maxwell's equations had such formally equivalent twins was already known in the nineteenth century, though the possibility that magnetic monopoles might have an independent existence was not taken seriously.[2] Dirac thought otherwise.

8.2 Dirac's Theory

In 1931, and in the very same paper where he first modified his "hole" theory so as to incorporate the production of positrons, Dirac embarked on an examination of the consequences of the existence of the twins to Maxwell's equations, now embedded in the formalism of quantum mechanics as informed by the Dirac Equation.

[1]We'd like to express our appreciation to the creators of this figure for having made it available under the Creative Commons CC0 Universal Public Domain Dedication.

[2]For the historical details and references see (Milton 2006, p. 1639), (Goldhaber and Trower 1990, p. 431), and (Adawi 1976).

The object of the present paper is to show that quantum mechanics does not really preclude the existence of isolated magnetic poles. On the contrary, the present formalism of quantum mechanics, when developed naturally without the imposition of arbitrary restrictions, leads inevitably to wave equations whose only physical interpretation is the motion of an electron in the field of a single pole. This new development requires *no change whatever* in the formalism when expressed in terms of abstract symbols denoting states and observables, but is merely a generalization of the possibilities of representation of these abstract symbols by wave functions and matrices. Under these circumstances one would be surprised if Nature had made no use of it (Dirac 1931, p. 62).

More particularly, the use to be made by Nature was to fix the size of the smallest charge.

[The present paper] will be concerned essentially, not with electrons and protons, but with the reason for the existence of the smallest electric charge. This smallest charge is known to exist experimentally and to have the value e given approximately by $hc/2\pi e^2 = 137$ [where h is Planck's constant and c is the speed of light]. The theory of this paper, while it looks at first as though it will give a theoretical value for e, is found when worked out to give a connection between the smallest electric charge and the smallest magnetic pole. It shows, in fact, a symmetry between electricity and magnetism quite foreign to current views (Dirac 1931, p. 62).

As demonstrated by Dirac, the connection was encapsulated in the equation $hc/2\pi e\mu_o = 2$, where μ_o was the strength of the magnetic pole which "means that the attractive force between two one-quantum poles of opposite sign is $(137/2)^2 = 4692\frac{1}{4}$ times that between electron and proton. This very large force may perhaps account for why poles of opposite sign have never yet been observed (p. 72)."[3] As later summarized by Dirac:

If one supposes that a particle with a single magnetic pole can exist and that it interacts with charged particles, the laws of quantum mechanics lead to the requirement that the electric charge be quantized – all charges must be integral multiples of a unit charge e connected with the pole strength g [the former μ_o] by the formula $eg. = hc/4\pi$. Since electric charges are known to be quantized and no reason for this has yet been proposed apart from the existence of magnetic poles, *we have a reason for taking magnetic poles seriously*. The fact that they have not yet been observed may be ascribed to the large value of the quantum of pole (Dirac 1948, p. 817, emphasis added).

Dirac's proposed solution to the problem of charge quantization has been very appealing to contemporary theoretical physicists. As Joseph Polchinski, a leading string theorist, remarked:

One of Dirac's remarkable discoveries was the connection between magnetic monopoles and charge quantization. Very early in the history of quantum theory, he recognized the important connection between geometry and quantum mechanics. Dirac showed that in the presence of a magnetic charge g, in order for the quantum mechanics of an electric charge e to be consistent one had to have $eg. = 2\pi n$, where n is an integer. Thus, the existence of even a single magnetic charge forces every electric charge to be a multiple of $2\pi/g$. From the highly precise electric charge quantization that is seen in nature, it is then tempting to infer that magnetic monopoles exist, and indeed Dirac did so.. .. (Polchinski 2004, p. 147).

[3]For a concise review of Dirac's derivation see (Preskill 1984, pp. 466–468).

Many other theorists share his view and monopoles have received top billing in Grand Unified Theories, string theory, and super-string theories. Given their popularity Polchinski has concluded that "the existence of magnetic monopoles seems like one of the safest bets that one can make about physics not yet seen (p. 144)." Granting the theoretical attractiveness of Dirac's account of the magnetic monopole, and the safe bet that they exist, we now ask what experimental efforts were made to capture them in the wild and thereby establish their existence. Have they revealed themselves in any way analogous to the discovery of the positron and omega minus?

8.3 Blas Cabrera and the Saint Valentine's Day Event

Attempts to experimentally detect either directly or indirectly Dirac's magnetic monopole were not undertaken until the 1950s. There were no less than 14 such attempts but none of any noteworthy success.[4] A significantly more promising approach than those previous employed was to send the monopole through a conducting loop and measure the induced current in the loop. A magnetic monopole passing through a coil of wire will induce a net current in the coil, and a change in the magnetic flux. This is not the case for a magnetic dipole or higher order magnetic pole, for which the net induced current is zero. This effect can be used as an unambiguous test for the presence of magnetic monopoles.

Blas Cabrera and his colleagues (henceforth Cabrera) decided to take up the challenge and coopted some recent technology to translate the abstract theoretical possibility into a tangible and effective experiment. He used a loop of superconducting wire connected to the superconducting input coil of a SQUID (superconducting quantum interference device) magnetometer which could be used to measure extremely small magnetic fields.[5] It is sensitive enough to measure fields as low as (5×10^{-18} T) with a few days of averaged measurements. For comparison, the Earth's magnetic field is between $(25–65) \times 10^{-6}$ T.

A magnetic monopole passing through such a loop of superconducting wire will produce a change in magnetic flux through the loop of $4\pi g = hc/e$, where g is the magnetic charge of the monopole, h is Planck's constant, c is the speed of light, and e is the charge of the electron. This is twice the flux quantum of superconductivity $\Phi_o = hc/2e$. In sum:

> Such a detector measures the moving particle's magnetic charge regardless of its velocity, mass, electric charge, or magnetic dipole moment.. .. In the general case, any trajectory of a magnetic charge g which passes through the ring will result in a flux-quanta change of 2, while one that misses the ring will produce no flux change (Cabrera 1982, p. 1378).

---

[4] For references and review see (Preskill 1984, pp 522–528) and (Goldhaber and Trower 1990, pp. 431–432). One such attempt (Price et al. 1975) did become infamous because of serious errors in calibration and the existence of a plausible, prosaic alternative explanation.

[5] In 1970 as reported in Alvarez et al. (1970, p. 701) and mentioned in Cabrera (1982, p. 1381 n1) a similar technique was used whereby "the sample was transported along a continuous path threading the windings of a coil [where] the coil was made of superconducting material.".

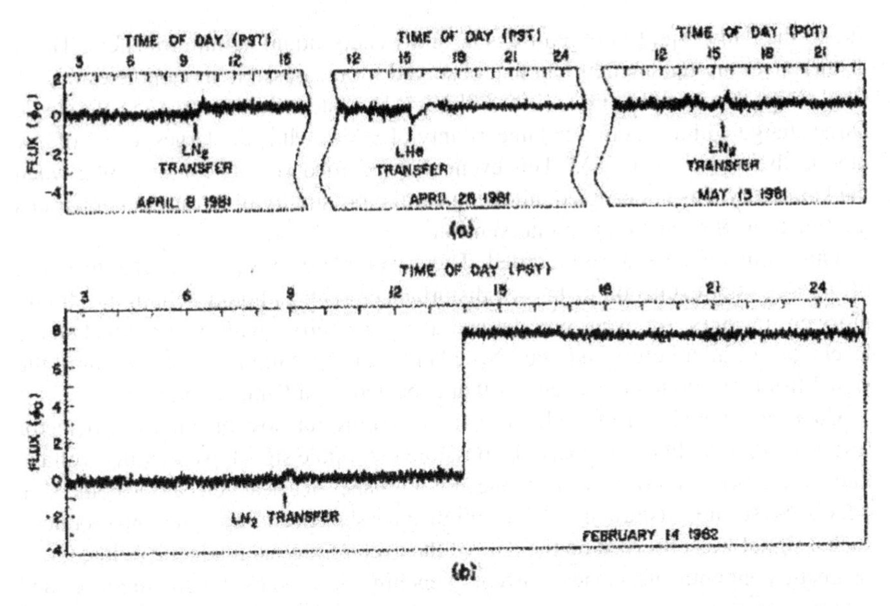

Fig. 8.2 Data records showing **a** typical stability and **b** the candidate monopole event. From Cabrera (1982)

This change in flux will cause a corresponding shift in the dc current level in the loop with a rise time of the order of the radius of the loop divided by the velocity of the monopole.[6] In order to maximize the sensitivity of the apparatus to flux change Cabrera's detector consisted of a four-turn 5 cm loop which meant that:

> The passage of a single Dirac charge through the loop would result in a change in flux of 8 Φ_o through the superconducting circuit, comprised of the detection loop and the SQUID input coil (a factor of 2 from $4\pi g = 2\,\Phi_o$ and a factor of 4 from the turns in the pickup loop) (p. 1379).

Because the magnetic monopole was expected to be relatively rare experimental runs were continuously conducted over a period of 151 days during which time "the candidate monopole event" was detected on February 14, 1982, Valentine's day.

The detector was calibrated in three different and independent ways: (1) by measuring the SQUID response to a known current in calibration Helmholtz coils ($\pm$4%), (2) by estimating the self-inductance of the superconducting circuit ($\pm$30%), and (3) by directly observing flux quantization within the superconducting circuit ($\pm$10%).[7] The calibrations agreed within the stated uncertainties. Figure [8.2] shows several intervals of data recording. There are typical small disturbances in the trace

---

[6]See (Preskill 1984, pp. 523–524) for further details on the analysis of the experiment.

[7]Later versions of the experimental apparatus would include calibration devices and techniques that would more closely approximate the signal due to a magnetic monopole. As we shall see, the fact that one could artificially produce such a signal cast some doubt on the reality of the initially observed event.

due to the daily liquid-nitrogen transfer and weekly liquid helium transfers. These disturbances are far smaller than that observed for the possible monopole event. "A single large event was recorded [Fig. 8.2b]. It is consistent with the passage of a single Dirac charge within a combined uncertainty of ±5%... It is the largest event of any kind in the record (p. 1379)." This event was recorded on 14 February 1982 when the laboratory was unoccupied, allowing for the possibility of a transient apparatus malfunction, or even a human intervention.

Other flux changes were recorded. There were 27 events exceeding a threshold of 0.2 Φ_o, after exclusion of known disturbances such as liquid helium and liquid nitrogen transfers. An event was defined as a sharp offset with well-defined stable levels for one hour before and after. No other event was within a factor of four of the signal from the single large event, or that expected for a Dirac monopole.

Cabrera devoted considerable effort to searching for possible spurious detector responses that might have caused the possible monopole signal. Neither line voltage fluctuations nor rf interference from the motor brushes of a heat gun caused detectable effects. No seismic disturbance, which might have shaken the apparatus and produced such a signal was observed on the date of the event. External magnetic fields, ferromagnetic contaminants, critical current quenching of the superconducting loop, and cosmic rays were also eliminated.

There was, however, one possible alternative explanation of the signal that Cabrera could not conclusively eliminate.

> Mechanically induced offsets have been intentionally generated and are probably caused by shifts of the four-turn loop-wire geometry which produce inductance changes. Sharp raps with a screwdriver handle against the detector assembly cause such offsets (Fig. 8.3). On two occasions out of 25 attempts these have exceeded 6Φ_0 (75% of the shift expected from one Dirac charge); however, drifts in the level were seen during the next hour (p. 1380).[8]

Cabrera was applying the Sherlock Holmes strategy of eliminating plausible alternative causes for the observation. Although he did not think that a mechanical effect was a likely cause of the observed signal, he did not feel that he could completely eliminate it as a possibility.

> A spontaneous and large external mechanical impulse is not seen as a possible cause for the event; however, the evidence presented by this event does not preclude the possibility of a spontaneous internal stress release mechanism. Regardless, to date the experiment has set an upper limit of 6.1×10^{-10} cm^{-2} s^{-1} sr^{-1} for the isotropic distribution of any moving particles with magnetic charge greater than 0.06 g (pp 1380–81).

In a later comment to Kent Staley, Cabrera remarked, "It was a striking event, because it was exactly the right step size [for a Dirac monopole], but I was not convinced because of the other possible although improbable mechanism (Staley 1999, p. 221)." Cabrera made no discovery claim for a magnetic monopole, but it remained a possibility.[9]

---

[8]This was unlike the observed signal, which remained constant for several hours (see Fig. 8.2).

[9]One might regard a signal that exactly fit the parameters for a Dirac monopole as evidence for the existence of the monopole. Cabrera felt, however, that his inability to eliminate the unlikely mechanical effect as a source of the signal argued against that conclusion.

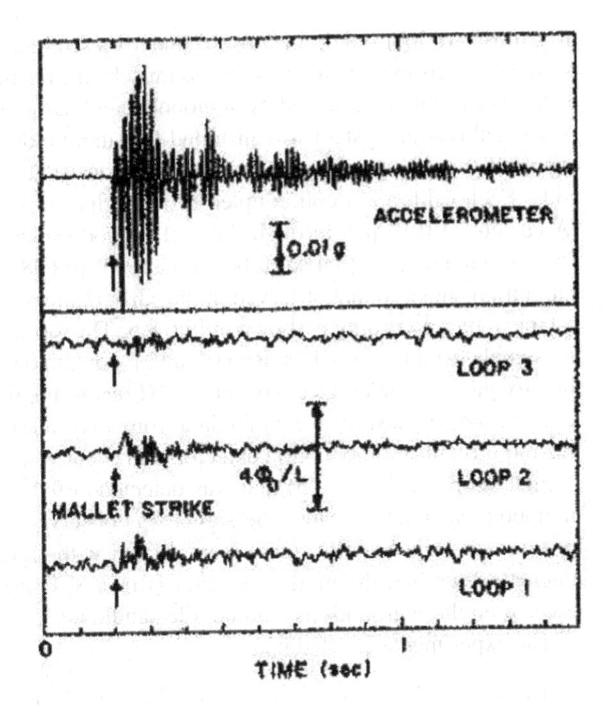

Fig. 8.3 Detector response to striking the detector with a mallet. From Gardner et al. (1991)

8.4 Further Work[10]

Cabrera and his collaborators continued their search for monopoles constructing larger and more sophisticated detectors. These were a three-loop detector and an eight-loop detector. These detectors had larger effective area than the initial detector, the eight-loop detector had an area of 1.1 m^2, and thus would be more likely to detect a monopole if it existed. These detectors also included more checks against effects that might mimic the signal expected for a monopole.

For the eight-loop detector these included

> This instrumentation includes a strain gauge attached to the exterior of the superconducting lead shield (to detect mechanical motion), a pressure transducer (to monitor the helium pressure above the liquid in the Dewar), and a power line monitor (to detect six different fault conditions). During most of our operating period a flux-gate magnetometer has been used to detect changes in the external field. We did not observe a significant correlation, so we have substituted a wideband rf [radiofrequency] voltmeter to detect changes in the local rf environment which can cause offsets in SQUID's. An ultrasonic motion detector monitors laboratory activity. When we perform activities known to disturb detector stability, we set a "veto" switch to prevent generating large numbers of useless computer events and to aid in calculating our live time (Huber Cabrera et al. 1990, p. 837).

To better isolate a real monopole signal they required a coincidence signal from two of the loops. "A feature of this geometry is that a monopole can induce a signal

[10]For more details and the history of Cabrera's later work see (Franklin 2016, Chapter 18).

in at most two loops and, for most of the cross section, no fewer than two loops. In contrast, offsets in more than two loops must be the result of electrical or mechanical disturbances and are rejected as monopole candidates (Huber et al. 1990, p. 835)." A new calibration system was installed that more closely approximated the signal expected for a magnetic monopole. These consisted of narrow, toroidally-wound coils. Each calibration coil coupled to two adjacent gradiometers simultaneously. "A current of 0.19 μA through the coil… produces a flux equivalent to that of a Dirac monopole (Fig [8.4]) (Huber et al. 1991, p. 638)." Note the similarity of the calibration signal to that observed in the early monopole candidate event (Fig. 8.2). A typical data recording is shown in Fig. 8.5. The top eight rows of the graphs show the signals from the SQUIDs. Rows S and F contain data from the strain gauge and the flux-gate magnetometer. Rows P and H record the pressure monitor and helium level sensor. The last rows contain data from a cosmic-ray channel (unused), power line monitor, ultrasonic motion detector, and event veto. Part (a) of the figure shows a single loop event (Loop 5) that was detected at 9:41 on 17 July 1987. It is not a monopole candidate because the signal was observed in only a single loop. Note, however, that the SQUID signal is correlated with signals in the strain gauge, the magnetometer, and the motion detector (Rows S, F, and U) and would have been rejected on these grounds as a monopole candidate.

The experimenters stated that

> In conclusion, these data set an upper limit of 7.2×10^{-13} cm^{-2} s^{-1} sr^{-1} at 90% confidence level on any uniform flux of magnetic monopoles passing through the Earth's surface at any velocity. This limit is a factor of 2000 below the flux suggested by the single-candidate event seen with the prototype detector. Based on this large factor and based on the noncoincident

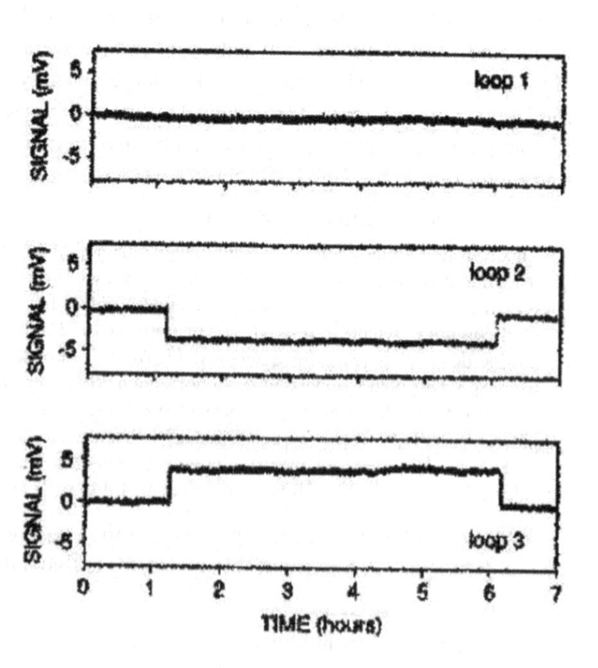

Fig. 8.4 Calibration signal equivalent to one Dirac charge in inductors Number 2 and 3. From Huber et al. (1991)

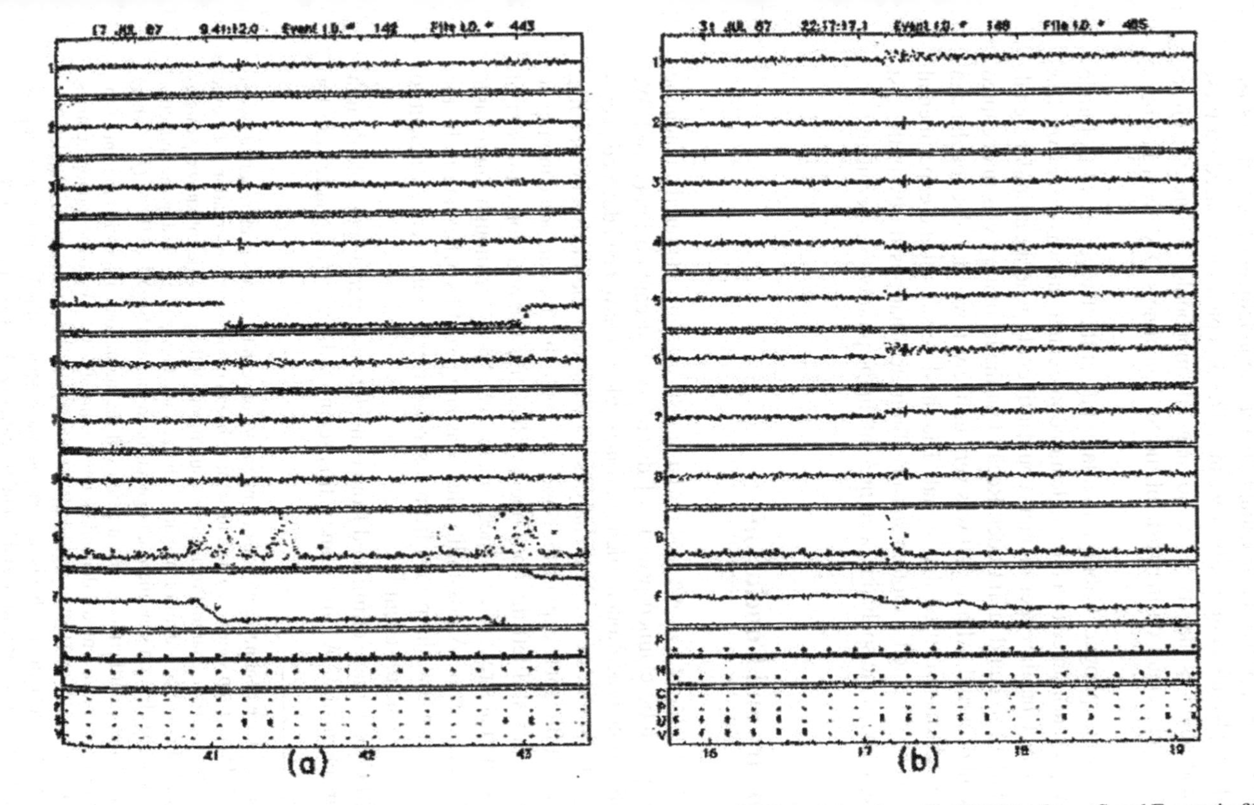

Fig. 8.5 Summary of an event data file for the eight-loop detector. The top eight rows contain filtered data from the SQUIDs. Rows S and F contain filtered data for the strain gauge and the flux gate magnetometer, respectively. Rows P and H contain unfiltered low frequency data from the pressure monitor and helium level sensor, respectively. The last row contains data from a cosmic ray channel (unused, power line monitor, ultrasonic motion detector, and event veto. **a** The event shown occurred at approximately 9:41. Note the coincidence between signals in the SQUID with signals in the strain gauge, magnetometer, and motion detector. **b** Event data demonstrating the resolution of Squid detectors. Note the oscillations in the SQUID output resulting from a sharp disturbance recorded in the strain gauge. From Huber et al. (1991)

nature of the prototype detector, we conclude that the entire data set from the prototype detector which contains the single event should be discarded (Huber et al. 1990, p. 838).

Once might have appeared to be enough, but in the judgment of both the experimenters and the physics community it wasn't. The accepted view based on these experiment and others is that there are no magnetic monopoles. Cabrera's refusal to make a discovery claim on the basis of the St. Valentine's Day event has been justified.

In an interesting epilogue, Cabrera, in a later talk at the University of Colorado (Franklin was present), stated that the single monopole candidate was made even less plausible by the fact that the experimenters could artificially generate a similar signal. If they could do it, so could an intruder. Recall that the laboratory was unoccupied when that event occurred. Although it was very unlikely, it was possible that human intervention might have caused that signal. The third version of the experimental apparatus, which contained a motion sensor, eliminated that possibility in the last run, and would have done so, had it been present, in the prototype experiment.[11]

8.5 Summary and Conclusion

Cabrera could have made a strong case for having successfully captured a magnetic monopole and thereby claimed priority of discovery. But caution intervened. We suspect that for some practitioners such caution would not have been deemed necessary because of Cabrera's extensive analysis showing that the likely confounding factors were not sufficient to explain the observed result, the "the candidate monopole event." Even so this still left the door open, at least for Cabrera, for some sort of "spontaneous internal stress release mechanism" as a "possible though improbable [interfering] mechanism." A possibility that was suggested by what had to be the surprising fact that a few sharp raps from a screwdriver handle were sufficient to closely duplicate the actual event.[12]

How could Cabrera have eliminated from consideration such a "possible though improbable mechanism"? At first glance, this seems to be a paradigmatic case for

---

[11] Although searches have continued, no convincing monopole candidate has been reported. As of 2020, the Particle Data group states, "To date there have been no confirmed observations of exotic particles possessing magnetic charge. Precision measurements of the properties of known particles have led to tight limits on the values of magnetic charge they may possess. Using the induction method (see below), the electron's magnetic charge has been found to be $Q_e^m < 10^{-24}\ Q_M^D$ (where Q_M^D is the Dirac charge). Furthermore, measurements of the anomalous magnetic moment of the muon have been used to place a model dependent lower limit of 120 GeV on the monopole mass. Nevertheless, guided mainly by Dirac's argument and the predicted existence of monopoles from spontaneous symmetry breaking mechanisms, searches have been routinely made for monopoles produced at accelerators, in cosmic rays, and bound in matter (Zyla, Barnett et al. 2020)".

[12] One has to wonder what might have happened if Cabrera had not thought to test the apparatus with something as unsophisticated as a screwdriver handle. Would there then have been an unqualified claim of discovery?

replication in the narrow sense of duplication. This because the event in question (the dramatic shift in measured inductance) might have been the result of accidental internal stress. So if no such event was detected in a duplicated experiment that would be good evidence that the first event was the result of such accidental (and non-recurring) internal stress, and moreover that the monopole had not been detected in the first experiment. On the other hand, if the event in question made a repeat appearance then you'd be back where you started, not knowing whether it was the registration of a monopole or just another manifestation of an accidental confounding cause. Thus, the duplication strategy makes sense only if you've good reason to think that the confounding cause won't occur a second time. But if it does (and you get the ambiguous event) then you will have wasted time and resources on the duplication.

There were two additional complications that Cabrera had to deal with. First, there was reason to believe that magnetic monopoles were not all that prevalent. On the basis of "hidden mass" estimates "as suggested from grand unification theories" Cabrera determined that the likely occurrence of a detectable monopole event would be "1.5 events per year through the detector loop (Cabrera 1982, p. 1381)." Second, there was a lack of specificity with respect to the worrisome "spontaneous internal stress release mechanism." One has to ask: caused by what? As Cabrera reported:

> Mechanically induced offsets have been intentionally generated and are probably *caused by shifts of the four-turn loop-wire geometry* which produce inductance changes. Sharp raps with a screwdriver handle against the detector assembly cause such offsets (p. 1380, emphasis added).

On the other hand, Cabrera's judgment, as we have seen, was that:

> A spontaneous and large external mechanical impulse is not seen as a possible cause for the event; however, the evidence presented by this single event does not preclude the possibility of a *spontaneous internal stress release mechanism* (p. 1380, emphasis added).

There's not much to work with here. Contrast this with the case of the Cox polarization experiment where the finger of blame pointed clearly in the direction of the Geiger counters which could be tested and replaced with more reliable measuring devices. Thus the problem for Cabrera was to obtain the *result* of generating a reliable signal from a magnetic monopole with an *improved* detection device that would reduce the possibility of interfering internal stress and make such stress independently detectable when it did occur. As described above, dealing with that problem was in large part the purpose of the three and the eight-loop detectors that were developed. In particular, to better isolate and identify a real monopole signal a coincidence in signal from two of the loops was required.

The strategy was successful in that confounding events were successfully segregated. But there was also an absence of the coincidence in signal that would have revealed the passage through the apparatus of a magnetic monopole. Thus, the melancholy conclusion that the earlier result had to be "discarded." Still, not a knock-out blow for the magnetic monopole given its rarity and its enthusiastic incorporation into Grand Unified Theories, string theory, and super-string theories making its existence "one of the safest bets that one can make about physics not yet seen" (Polchinski

2004, p. 144). Thus, despite the evident difficulty of detecting the magnetic monopole it has been deemed at least by some experimentalists to be worth the effort to continue the search.[13]

References

Adawi, I. 1976. Thomson's monopoles. *American Journal of Physics* 44: 762–765.

Alvarez, L.W., P.H. Eberhard, et al. 1970. Search for magnetic monopoles in the lunar sample. *Science* 167: 701–703.

Cabrera, B. 1982. First results from a superconductive detector. *Physical Review Letters* 48: 1378–1381.

Dirac, P.A.M. 1931. Quantised singularities in the electromagnetic field. *Proceedings of the Royal Society (London)* 133 (821): 60–72.

Dirac, P.A.M. 1948. The theory of magnetic poles. *Physical Review* 74: 817–830.

Franklin, A. 2016. *What makes a good experiment?*. Pittsburgh: University of Pittsburgh Press.

Gardner, R.D., B. Cabrera, et al. 1991. Search for cosmic-ray monopoles using a three-loop superconductive detector. *Physical Review D* 44: 622–635.

Goldhaber, A.S., and W.P. Trower. 1990. Resource letter MM-1: Magnetic monopoles 58. *American Journal of Physics* 58: 429–439.

Huber, M.E., B. Cabrera, et al. 1990. Limit on flux of cosmic-ray monopoles from operations of an eight-loop superconducting detector. *Physical Review Letters* 64: 835–838.

Huber, M.E., B. Cabrera, et al. 1991. Search for a flux of cosmic-ray monopoles with an eight-channel superconducitng detector. *Physical Review D* 44: 636–660.

Milton, K.A. 2006. Theoretical and experimental status of magnetic monopoles. *Reports on Progress in Physics* 69: 1637–1711.

Polchinski, J. 2004. Monopoles, duality, and string theory: 145–154. *International Journal of Modern Physics A* 19 (supp01): 145–154.

Preskill, J. 1984. Magnetic monopoles. *Annual Review of Nuclear and Particle Science* 34: 461–530.

Price, P.B., E.K. Shirk, et al. 1975. Evidence for detection of a moving magnetic monopole. *Physical Review Letters* 35: 487–490.

Staley, K. 1999. Golden events and statistics: What's wrong with Galison's image/logic distinction? *Perspectives on Science* 7: 196–230.

Zyla, P.A., R.M. Barnett, et al. 2020. Review of particle physics. *Progress of Theoretical and Experimental Physics* 2020: 083C001 (to be published).

[13]For reviews of recent experimental searches see (Milton 2006, pp. 1690–1705) and (Zyla et al. 2020).

Chapter 9
Conclusion

In this chapter we aim to both summarize and refine the lessons to be learned from our case studies. More specifically, we'll highlight the various concepts we've employed to understand the historical practice and show how they apply in different ways to the cases studied where those differences depend—not surprisingly—on the specifics of the experimental activities under review.

9.1 Starting Points

As we noted in our Introductory chapter, that the results of a scientific experiment must be reproducible has been taken to be axiomatic and as such the "Gold Standard" for scientific objectivity. At first glance, this seems obvious. So why make a fuss? After all, consider what would happen if you repeated an experiment and got a different result. You'd reconsider that publication submission and, if necessary, have the presses shut down.

But does it follow that what the Gold Standard requires, as a necessary condition for scientific objectivity, is that any experiment result must in fact be reproduced using the same experimental apparatus and procedures? That goes too far if only because such a repetition would only serve to reproduce the experimental deficiencies, if such there be, that existed in the original experiment. To be fair, however, such a more or less exact repetition might under the right circumstances reveal the existence of a confounding cause that varied during the course of the experiment. This, for example, was the case in the Cox experiment where the data varied among the experimental runs indicating the presence of a varying confounding factor.

At bottom, much of the concern about the need for scientific objectivity comes from the worry that an experimental result may reflect the influence of confounding factors rather than the underlying fundamental processes that the experiment aims to uncover. Making use of the current terms of art, the problem is how to deal with *systematic uncertainty*. It's clear as shown by our case studies that what's needed are

A. Franklin and R. Laymon, *Once Can Be Enough*,
https://doi.org/10.1007/978-3-030-62565-8_9

effective ways of reducing such uncertainty by means of improved experimental apparatus and procedures. Thus, while there may be no harm in duplicating an experiment it's much better to think more deeply and aim for the better experiment.

There is an ambiguity in all this that needs to be dealt with and that concerns the notion of the *result* of an experiment. So, for example, does the Gold Standard just deal with the data produced, such as density and time determinations or cloud chamber photographs, or should it be understood more expansively to include higher level theoretical processes and entities such as the replication of DNA or the existence of the positron? Does the Gold Standard require then that every such theoretical result be retested using the same apparatus and procedures or will different apparatus and procedures suffice? If the latter, then the Gold Standard is just the truism that scientific results at every level are open to doubt and should be continuously put to the test by whatever form of experiment seems most efficacious given concern about systematic uncertainty.

Our case studies were undertaken with the aim of discovering *methodologically relevant specificity* regarding the replication of experimental results. We now take the opportunity to review what our cases have revealed.

9.2 Interaction of Theory and Experiment

Experimental investigations may be undertaken in response to a determinative theoretical prediction, in which case their purpose is to test that prediction, or they may have been undertaken in an effort to search for answers to questions that arise because of an absence or incompleteness of applicable theory. In the latter case they take on the character of being exploratory. The distinction is important for determining what constitutes the result of an experiment and whether and in what ways the experiment should be replicated. Our cases exemplify both sorts of experimental response.

Consider the Meselson-Stahl experiment. The assumed background for the experiment starts with the assumption that the Watson-Crick double helix correctly describes the components and structure of a DNA molecule. If so, then given what was suspected to be the role of DNA in genetic inheritance, the question was how does it replicate? There was agreement that there were likely to be only two such answers, namely, as explained in an earlier chapter, conservatively or semi-conservatively.[1] There was no theoretical answer to this question. Meselson and Stahl answered the question with their brilliantly conceived combination of centrifuge and hungry DNA gobbling up nutrient that was spiked with nitrogen isotopes of different weight that would correspond to different stages of the replication. The question of semi- versus fully conservative replication was thereby answered (without commitment as to the

[1] There was in addition the further question whether the DNA helixes replicated intact or whether as proposed by Delbruck (as a way of solving the "unwinding" problem) they split up into smaller units which then recombined in a way that mimicked either conservative or semi-conservative replication. To answer this question, Meselson and Stahl developed a different experimental test.

number of DNA molecules in a replicating unit). In addition, Watson-Crick's favored position that a single DNA molecule replicated semi-conservatively was confirmed in the sense of being consistent with the experimental "subunits" result. So, while the motivation may have been to test the Watson-Crick proposal, Meselson and Stahl were careful not to restrict the applicability of their results to just that.

The discovery of the positron is more straightforward. Anderson made clear that he knew little of Dirac's theory and that he came upon the discovery only after an arduous process of trying to make sense out of the cosmic ray photographs. Moreover, the initial motivation of his experimental research was not to discover or confirm the existence of a new sort of particle but rather to engage in a more prosaic form of cosmic ray research. Here the principal theoretical components used to interpret the ionization trails dealt with the production of such trails and most importantly with their length before stopping. Blackett could not claim that he did not know of Dirac's theory, but he was scrupulous in avoiding any use or reference to it until after he had nailed down the discovery and identification of the positron. Only then did he make use of Dirac's theory in explaining the production of positrons. So, while Dirac made the prediction, the discovery of the positron was not motivated by the prediction nor was Dirac's theory employed or needed for its discovery.

It might be claimed that the then current belief that the only charged particles were electrons and protons obviously entails the prediction that there would be nothing else. And that the discovery of the positron therefore constituted a disconfirmation of that prediction. It's certainly true that Anderson in particular found himself hard pressed to deny that prediction and only as a sort of last resort felt compelled—due in part to the "golden" photograph—to assert the existence of the new particle. Still, the fact remains that searching for such disconfirmation was not the original research purpose for either Anderson or Blackett.

The discovery of the Omega minus particle comfortably fits the pattern of an experimental search made for the purpose of confirming or disconfirming a theoretical prediction. But here there was the novelty that the Eightfold way was a phenomenological theory which made use of certain symmetries in the strong and weak forces to restrict the group theoretical structures that newly discovered particles could occupy. One of these structures, the decuplet, after being partially filled by known particles became successively inhabited by newly discovered particles leaving only one slot needing to be filled. Gell-Mann's prediction was that a suitable particle for residency, dubbed the "Omega minus", would be discovered and thus made available to occupy the remaining slot. And with the discovery of the Omega minus, the Eightfold way achieved respectability to the point of justifying further research as to its underlying dynamic basis.

Mendel's efforts were first and foremost to develop an exploratory, experimental strategy that would uncover whatever regularities there might be in the inheritance of plant traits. The strategy was successful because his experiments on pea plants not only suggested what became known as the laws of segregation and independent assortment, but also provided strong evidence for those laws—which in turn encouraged the application of the strategy to other species. In addition, as in the case of the

Eightfold way, it set the stage for further research regarding the underlying dynamics based on, for example, plant chromosomes, of these patterns.

In the case of the Cox and Chase experiments, a theoretical prediction was involved from the very start but only in an attenuated sense. Cox recognized that given de Broglie's wave proposal there was an analogy to be made between the spinning electron and the electric and magnetic fields of ordinary light—and moreover that this analogy could serve as the basis for an experimental test for the existence of electron polarization. Furthermore, that such an experiment could be developed to function in a way analogous to that of Barkla's experimental determination that X-rays could be polarized. Hence, there was a theoretical prediction but only analogy supported the prediction. The anticipated asymmetries were detected but Mott preempted any celebration by coming up with a genuine theory based on the Dirac equation that predicted asymmetries but located at different orientations of the apparatus. This led to a period of great uncertainty that lasted for many years and was not fully resolved until the proposal by Lee and Yang that parity was not conserved and the experimental confirmation of that proposal by Wu and others. And here we note that the proposal by Lee and Yang was not a theory derived prediction but rather a proposal that if accepted would resolve the $\theta - \tau$ puzzle.

The relationship between Cabrera's experiments and Dirac's theoretically based prediction of the existence of a magnetic monopole falls cleanly in the category where experiments are developed from the start with the purpose of testing a theory-based prediction. Moreover, unlike the cloud chambers used by Anderson and Blackett, Cabrera's experimental apparatus were specifically designed and optimized for the registration of a passing monopole.[2] But unlike our other cases, Cabrera's experimental tests failed to confirm the prediction—but did leave the door open for others to search for the elusive monopole.

In sum, our case studies illustrate the differences between experiments undertaken in response to a theoretical prediction as opposed to those undertaken in an effort to search for answers to questions raised because of an absence or incompleteness of applicable theory.

9.3 Experiments as Complex Events

When is the replication of *an experiment* warranted? Presumably, a sensible question. But to describe the question this way encourages thinking of an experiment as simply

[2]It can happen when testing contrary predictions of competing theories that the optimal apparatus for one will be quite different than that for the other. In such cases, theory comparison is seriously compromised. For example, using an interferometer to measure etherial velocity requires (for the type of ether that Dayton Miller was proposing) that the interferometer not be significantly insulated whereas if one wants to reduce the confounding effect of changing temperature gradients then a well-insulated interferometer is required. See Franklin and Laymon (2019, 8.4–8.9) for how Miller systematically expanded the operational range (in terms of orbital positions at which measurements were taken) of his interferometer in order to circumvent this sort of incompatibility.

turning on the apparatus start switch and then waiting for the results to print out. And where on this approach, replication means simply pressing the start switch again. But as shown by the many examples we've discussed what's often referred to as an experiment is in actual practice better described as being a coordinated set of events, procedures and apparatus that taken together constitute *the* experiment.

Consider, for example, our first case study, the Meselson-Stahl experiment. There was, first of all, a demonstration of viability of the procedure by showing that centrifuging DNA in a concentrated solution of cesium chloride resulted in a stable concentration gradient. Second, after the experimental runs were completed it was determined that the locations of the various density peaks and dips lined up coherently. All of this constituted a *calibration* process that demonstrated that the experimental procedure was capable of registering the different stages in the DNA replication process in a way that would lead to quantitatively coherent data. Finally, as an exercise of caution Meselson also conducted an additional experimental run which extended the DNA replication process through several more generations. In short, he replicated his own experiment and then reported it, somewhat misleadingly, as if it consisted of a single experiment run.

Anderson's "golden" photograph did not make its appearance by itself. It came with several of its siblings that were intended to highlight additional features of the positron—such as its association with particle track showers. And for both Anderson and Blackett multiple shower photographs were needed to justify the probabilistic conclusion as to the likely origin of a cloud chamber track. There was in addition much in the way of initial refinement and tuning of the apparatus—although as is typical of experimental practice most of that never gets reported. It's part of what is implicitly understood as being part of good experimental practice.

Much, though not all, of the complexity in the make-up of an experiment is a built-in attempt to deal with *systematic uncertainty*. But by so doing one wards off—at least in part—demands for replication because of the possibility of unaccounted for confounding causes. This because by design the experiment has already taken account of the replications needed to do so. The Cox and Chase electron polarization experiments provide a particularly clear case of such a built-in attempt to deal with systematic uncertainty. As the reader may recall, one problematic aspect of Cox's original data was that although the average of the observed asymmetries was consistent with polarization there was variation in the direction of the asymmetries observed in of experimental runs. This suggested that there was a confounding cause at work that varied during the experimental runs. The response was to separately test the consistency of the Geiger counter used and then replace it with a more consistent electroscope. This sort of experimentation on and consequent modification of the apparatus should, we submit, be counted as a constituent component of the experiment. Moreover, if there be any doubt on that we note that this subsidiary experimentation was reported in Chase's published account of the experiment. Similarly, Anderson in his 1933 *Physical Review* announcement of the discovery of the positron very carefully dealt with and eliminated alternative explanations. There was also a careful consideration of the errors inherent in the determination of ionization levels.

We could go on in a similar vein and show that close analogues of the procedures described above exist for all of the cases we have examined in the preceding chapters. Assuming that this is so, the obvious conclusion is that a scientific experiment is best conceived as a complex assemblage of apparatus and procedures many of which are meant to test the reliability and efficacy of the apparatus itself. What though is the relevance of viewing scientific experiments in this way when it comes to questions of the need for replication? Answer: When considering the need for replication, you have to ask *what exactly is it that warrants replication*? *Which of the many components of an experiment warrants replication*? *All or only some of them*? Making headway on answering such a question requires determining what constitutes the *result* of an experiment. The idea here is that it is the *result* that warrants replication. But since, as we shall see, there are many candidates for what constitutes the result there will be correspondingly many candidates for what it is that warrants replication.

9.4 What Constitutes the Result of an Experiment?

An experiment may be understood, depending how you look at it, as having several different results. At the one extreme there is the "raw data" result while at the other end are the higher-level theoretical consequences. With respect to replication there is the obvious direct connection whereby if you were to replicate the raw data result by means of a duplicate experiment you would automatically replicate the theoretical result as well. But there's an important indirect connection to be kept in mind. Using a different sort of experiment you could seek to replicate the higher level result and in this way indirectly substantiate the raw data result of the initial experiment. That is, a successful replication of the theoretical result by means of a different experiment serves as evidence that the raw data result of the original experiment did in fact reflect the underlying processes that the experiment was designed to detect. Similarly, a failure to replicate the theoretical result casts doubt on the raw data result of the original experiment—especially if the second experiment is arguably better designed.

The Meselson and Stahl experiment is a clear example of what we have in mind. Here there were different levels of theoretical attachment that required different appraisals as to what constituted the result of the experiment. There was first of all what we had described in an earlier chapter as the "raw data" results—which in this case were concisely reported and displayed in the photographs and graphs contained the Meselson-Stahl (henceforth Meselson) published report. To describe such data as "raw" means only that the theoretical presuppositions involved were not at issue and were relevant to the operation of the apparatus and not dependent on the alternative hypotheses at issue: semi-conservative versus conservative replication.

Next in the hierarchy is the set of what Meselson referred to as the "subunit" conclusions which constituted a higher-level theoretical result. Here the idea was to use the raw data as premises in an argument that would deductively entail the replication properties of the subunits. But as we noted Meselson's deduction required the assumption that the subunits replicated once and only once in each generation time. Finally, and what was in fact the original aim of the experiment, there was the result of having confirmed that if the molecule responsible for the transfer of genetic information consisted of a single Watson-Crick double helix then its replication was semi-conservative and not conservative.

Thus, at the lowest level one might insist on a duplication of the experiment in order to replicate the raw data result. In fact, as we noted earlier, Meselson more or less did exactly this when he conducted an additional experimental run that extended the replication process into additional generations. But because once was historically deemed enough for the experiment as reported, there was no call that the raw data results be further replicated or more importantly (since confounding errors might simply be repeated in a replication of the raw data) that the subunits result or the confirmation of the Watson-Crick proposal be replicated in some other way.

Blackett's procedure is somewhat analogous to that employed by Meselson in that the bulk of his 1933 report makes no mention of Dirac's hole theory and its confirmation by the experimental results. The principal theoretical result for Blackett was the discovery of the positron where the required theory necessary for analyzing the raw data to get this result was carefully delineated. In the case of Meselson, the analogue was the subunits result. Only after the existence of the positron was determined to have been firmly established did Blackett bring Dirac into the picture—and this in a twofold way. First, there was the confirmation of Dirac's prediction of the existence of the positron. Second, as discussed in our chapter on the positron, Blackett made use of Dirac's theory to support his proposed analysis of the production of positrons. Hence, there was this further result about what were likely to be the production and annihilation "mechanisms" for positrons.

We next want next to focus our attention on the question of for what *purpose* was once enough in these and the other cases we've examined.

9.5 Once Was Enough, but for What Purpose?

We've relied on the expression "once was enough" as a short hand for our claim that there was little if any point in replicating an identified result of an experiment—where the result could be the raw data or that involving a higher level of theoretical involvement. But as we've suggested in our case studies, any judgment that once is enough must take into account a consideration of the question of for what *purpose* is once enough. This because the reliability and range of an experimental result will constrain what purpose or purposes can be served by the result.

By way of better understanding such constraints we have appealed to a distinction between two sorts of purpose broadly conceived. First, that once is enough for accepting a result as well confirmed and thus unlikely to have that status easily changed. Second, that once is enough for accepting the result as worthy of further investigation and development even though it does not fall into the first category. In a nutshell, the distinction is between *acceptance* and *pursuit*.[3] Obviously, this division is not clear cut since as the history of science demonstrates even well confirmed and accepted results have been overturned, and pursuit if broadly construed will open the doors for virtually all results.[4] There is also the fact that well confirmed results will serve as the foundation for further investigation and development. Still, our case studies serve as suggestive paradigms useful for further development of the distinction.

In the case of the Meselson experiment once was enough for the results of that experiment to gain a well-deserved admission into the class of well confirmed results. Once was also clearly enough for the 1933 experiments of Anderson, and Blackett and Occhialini (henceforth Blackett) to have established the existence of the positron. We think that once was also enough for Anderson's experiment considered by itself. Their discovery of the positron obviously justified the *pursuit* of more accurate determinations of the positron's mass and charge. Once was also enough for the Anderson and Blackett experiments, either singly or in conjunction, to have justified the *pursuit* of experimental data and theoretical development relevant to the *production* of positrons by gamma rays including, for example, the pair-wise theory of positron creation proposed by Blackett. Because of its many exotic and problematic features, Blackett argued that once was not enough to have confirmed Dirac's hole theory in anything other than the provisional and qualified sense of having shown the applicability of Dirac's calculation for the life expectancy of the positron.

Unlike the discovery of the positron which was somewhat accidental and unanticipated, the search for the Omega minus particle was expressly made with the intent to discover such a particle and thereby confirm the prediction of its existence by the Eightfold Way. And as described in an earlier chapter, there were two sightings (using different decay processes) reported in quick succession by the Brookhaven National Laboratory. As a result, there emerged an agreement that once was enough to conclude that the Ω^- particle existed which in turn justified the pursuit of its spin and parity as well as a more accurate determination of its mass. Also justified was the pursuit, on the basis of the Eightfold Way, of a physics that would yield the so far missing underlying dynamics.

Mendel's experiments serve to illustrate the role played by the determination that once was enough for *pursuit* even if not enough for complete *acceptance*. While somewhat muddled, it is fair to say that the overriding intent of the experimental endeavors made after the rediscovery of Mendel's work was not so much to test Mendel by replication but rather to elaborate and expand on his methodology. Thus, at the least, once was enough for Mendell's pea experiments to have justified the pursuit and further development of his methodology and its consequences for what ultimately became known as genetics. In short, what mattered historically was not

---

[3]As we indicated in an earlier chapter, this distinction was first proposed in (Laudan 1977, pp. 100–114), questioned in (Nickles 1981, pp. 100–108), and applied to certain episodes in particle physics in (Franklin 1993).

[4]Thus, we note the somewhat surprising corollary that a call for replication indicates that the experiment was worthy of pursuit if only in the form of replication.

so much the exactitude of Mendell's results as the role played by his methodology as a template for such further development. Understood this way, whatever defects that may have existed, for example, in Mendel's data collection were subsumed in the process of the further successful development and elaboration of Mendel's methodology and the corresponding laws of inheritance.

The many twists and turns and changes of fortune for the Cox and Chase experiments provide a challenging problem for determining whether once was enough and if so for what purpose—a problem for both the historical participants as well as for later day commentators. As we explained in an earlier chapter, Chase successfully refined the apparatus so to as convincingly demonstrate the raw data result of an asymmetry at the 90°/270° orientations and less convincingly so at the 0°/180° orientations—from which followed the theoretical result of electron polarization. Thus, tantalizingly close for the purpose of *acceptance* and surely, one would have thought, for the purpose of *pursuit* (in terms of extensions and refinement) of the Chase result. But these appraisals were rendered suspect because of Mott's contrary theoretical determination that the greatest asymmetry was to be found at the 0°/180° orientations and that the scattering was symmetrical at the 90°/270° orientations. But given the preponderance of the later reported null results for the 0°/180° asymmetry, pursuit came to be restricted to the conflict of those null results with Mott's theory. In the process Chase's positive result for the 90°/270° asymmetry ceased to be of interest *both* as a matter of acceptance and of pursuit. And so, it lay dormant until the discovery of parity nonconservation after which it was realized that Cox and Chase had unknowingly produced the first experiments that provided evidence for the nonconservation of parity. This case thus nicely illustrates the uncertainty that may exist when it comes to ascertaining the theoretical result of an experiment. Here Cox and Chase with nothing more in hand than an analogy between de Broglie's wave account and the electric and magnetic fields of ordinary light assumed, not unreasonably, that like X-rays beta rays arrived unpolarized and could be polarized with an experimental apparatus analogous to that use by Barkla to polarize X-rays.

The experiment by Wu and her colleagues provides an informative variation on the interactions between experimental result, purpose served and whether once was enough. For in that case even though once was enough (or so we argued) for the experiment to have established the result that parity was not conserved, there were independent attempts by Garwin and his colleagues, and Friedman and Telegdi to replicate that result using entirely different sorts of apparatus and procedures. Since all three experiments led to the same result—parity nonconservation—where that commonality served to indirectly confirm the raw data results of each of the experiments.

In the case of the first of Cabrera's monopole experiments once was deemed not enough by Cabrera himself—although the case was arguably a close call. So in the experiments that followed the aim was to reproduce the (type of) signal that was detected in the first experiment but in a way that significantly reduced the possibility of interfering internal stress and would make such stress independently detectable when it did occur. In other words, *to essentially replicate the raw data of the first experiment* but in a way that made its interpretation as that of a monopole not subject to doubt. While he was successful in separating out the confounding signal there was no monopole signal.

9.6 Concluding Remarks

The Gold Standard that the results of a scientific experiment must be reproducible appears at first to be both stern and of singular axiomatic significance. Not so we submit. To begin, it is not a necessary condition that the raw data result of an experiment be reproduced in order to pass scientific muster. Moreover, attempting to reproduce the raw data by using identical or nearly identical experimental apparatus and procedures is the least effective way of removing whatever uncertainty may exist about such data. At best it may reveal the occurrence of varying confounding factor. As amply shown by our historical examples, it is much, much better by far to replicate not the raw data but rather the higher level experimental result by means of experimental apparatus and procedures designed specifically to deal with the uncertainties involved in the original experiment.

Pushing this strategy a step further is to replicate the higher level theoretical results of an experiment using an entirely different sort of apparatus and procedures. In this way such replication serves as evidence that the original experiment had been effectively executed and analyzed and by so doing arguably satisfies, though indirectly, the reproducibility requirement of the Gold Standard. To replicate a result in this indirect way without having to reproduce the original experiment obviously opens the door to a wide range of experimental possibilities.[5] Consider, for example, this sequence of replications: (Cox et al. 1928; Chase 1930a, b; Frauenfelder et al. 1957; De Shalit et al. 1957), where the result sought to be replicated was that beta rays arrived on the scene not unpolarized but rather as longitudinally polarized. If one takes the aim of replication to be that of replicating the result that parity is not conserved in weak interactions then the experimental sequence can be expanded to include (Wu et al. 1957; Garwin et al. 1957), and (Friedman and Telegdi 1957). Once, however, the constraints on replication are relaxed this way the Gold Standard, so understood, merges into the desideratum that experimental results should be *independently testable*, i.e., testable in ways that rely on different underlying theoretical and operational suppositions.[6] Still while desirable, not necessary because, as we have shown, in some cases *once is enough*. We do not mean to be claiming that once will always, no matter what, be understood as having been enough only that all foreseeable arguments for assuming otherwise have been exhausted. This goes as well when the purpose of once being enough is that of justifying pursuit as opposed to acceptance.

[5] As shown in (Franklin 2018) it will often be difficult to determine whether the same result has been achieved given differences in background assumptions and procedure.

[6] We will not attempt here to further analyze what exactly this desideratum comes to. For an inkling of what's involved see Laymon (1980).

In sum, our aim has been twofold, to excavate and further analyze the historical argumentation used to support appraisals that once was enough, and then to deal with the apparent conflict between the Gold Standard and judgments that once was enough.

References

Chase, C.T. 1930a. The scattering of fast electrons by metals I. *Physical Review* 36: 984–987.

Chase, C.T. 1930b. The scattering of fast electrons by metals II. *Physical Review* 36: 1060–1065.

Cox, R.T., C.G. McIlwraith, et al. 1928. Apparent evidence of polarization in a beam of b-rays. *Proceedings of the National Academy of Sciences (USA)* 14: 544–549.

De Shalit, A., S. Cuperman, et al. 1957. Detection of electron polarization by double scattering. *Physical Review* 107: 1459–1460.

Franklin, A. 1993. Discovery, pursuit, and justification. *Perspectives on Science* 1: 252–284.

Franklin, A. 2018. *Is it the same result? Replication in physics.* San Rafael, CA: Morgan and Claypool.

Franklin, A., and R. Laymon. 2019. *Measuring nothing, Repeatedly*. San Rafael, CA: Morgan and Claypool.

Frauenfelder, H., R. Bobone, et al. 1957. Parity and the polarization of electrons from Co^{60}. *Physical Review* 106: 386–387.

Friedman, J.L., and V.L. Telegdi. 1957. Nuclear emulsion evidence for parity nonconservation in the decay chain $\pi^{+} - \mu^{+} - e^{+}$. *Physical Review* 105: 1681–1682.

Garwin, R.L., L.M. Lederman, et al. 1957. Observation of the failure of conservation of parity and charge conjugation in Meson decays: The magnetic moment of the free Muon. *Physical Review* 105: 1415–1417.

Laudan, L. 1977. *Progress and its problems: Toward a theory of scientific growth.* Berkeley: University of California Press.

Laymon, R. 1980. Independent testability: The Michelson-Morley and Kennedy-Thorndike experiments philosophy of science. *Philosophy of Science* 47 (1): 1–37.

Nickles, T. 1981. What is a problem that we may solve it? *Synthese* 47: 85–118.

Wu, C.S., E. Ambler, et al. 1957. Experimental test of parity nonconservation in beta decay. *Physical Review* 105: 1413–1415.